The Logic of Miracles

THE LOGIC
OF MIRACLES

MAKING SENSE OF RARE,
REALLY RARE, AND
IMPOSSIBLY RARE EVENTS

LÁSZLÓ MÉRŐ

TRANSLATED FROM THE HUNGARIAN
BY MÁRTON MOLDOVÁN

TRANSLATION EDITED BY DAVID KRAMER

Yale
UNIVERSITY PRESS
New Haven and London

Published with assistance from the foundation established in memory of Amasa Stone Mather of the Class of 1907, Yale College.

Yale University Press books may be purchased in quantity for educational, business, or promotional use. For information, please e-mail sales.press@yale.edu (U.S. office) or sales@yaleup.co.uk (U.K. office).

Set in Adobe Garamond and Gotham types by Newgen North America. Printed in the United States of America.

Library of Congress Control Number: 2017955615

ISBN 978-0-300-22415-3 (hardcover : alk. paper)

A catalogue record for this book is available from the British Library.

This paper meets the requirements of ANSI/NISO z39.48-1992 (Permanence of Paper).

10 9 8 7 6 5 4 3 2 1

Contents

CONTENTS

Preface

This book is about secular "miracles"—highly unusual events—
that have profoundly changed the world economy, forcing us to
change how we think and educate ourselves. Modern mathematics
can help us understand how secular miracles work, how positive
miracles have created a mechanism for continuous growth—I call it
the rich man's junk heap—that helps us to right ourselves in the after-
math of crises caused by negative miracles. Crisis and growth spring
from a common source.

The impulse to write this book came when I read Nassim Nicholas Taleb's bestseller *The Black Swan*. Taleb calls the world of everyday events Mediocristan and the world of unusual events—especially negative events, which he calls black swans—Extremistan. He
maintains that the models that describe Mediocristan are obsolete.

I was impressed by Taleb's interesting and sometimes profound
ideas but repelled at the same time. First of all, despite my advanced
degree in mathematics, I found his mathematical presentation completely incomprehensible. Perhaps not coincidentally, I also found
his conclusions implausible. As the author of several textbooks and
books on popular science, I set out to write a book that would both
challenge some of Taleb's conclusions and present the mathematical concepts in a way that non-mathematically-minded readers could
understand.

While I was writing the book, however, my thinking changed in two major ways. First, although the mathematics of Extremistan is useful for describing certain phenomena, I found that the old mathematics of Mediocristan is hardly obsolete. The name "Mediocristan" is too pejorative, so I renamed these regions Wildovia and Mildovia—the wild world and the mild world. Miracles (unusual events) occur in both places, but they obey different mathematical laws in each. My book grants equal time to the mathematics of Mildovia and to that of Wildovia.

Second, I realized that although the features of Wildovia—stock market crashes, earthquakes, wars—are common in our modern world, we should live our lives and organize our economies as much as possible according to the laws of Mildovia. A Wildovian economic model might be more precise, but Mildovian models serve us better. Both models, Mildovian normalcy and Wildovian turbulence, feed upon themselves. When we believe in a Mildovian model, we actually get a more Mildovian economy and society, whereas the psychological effect of believing in a Wildovian model would lead to political and economic chaos.

I believe that we are best served by a Mildovian model of human nature in which we view ourselves as generally mild-mannered and civilized and in which we see war, criminality, and rioting as extraordinary occurrences. Nevertheless, wars and other calamities occur with a frequency that can be predicted by a Wildovian model. Perhaps, despite being a Wildovian species, we are able to maintain a somewhat civilized society only because we believe we *are* civilized and therefore use a Mildovian model to govern our lives.

How, then, do we learn to live with the fact that crises are bound to occur sooner or later without our being constantly in crisis mode? We must prepare for crises so that when they hit, they will not fatally harm us. First of all, we have to learn how to model both Mildovia's and Wildovia's phenomena. We require one model for normally occurring events, some of which will be relatively uncommon, and another for extremely rare, extraordinary events. Features of these two radically different worlds are simultaneously present in our lives.

There is a certain amount of mathematics involved in this book, and if you are able to follow it, it will enrich your understanding of what I have to say. But if you were hoping to see no mathematics whatsoever and yet are eager to learn about the nature of miracles, feel free to skip the equations and concentrate on the general concepts. If a little mathematics doesn't frighten you but you don't want to delve too deeply into it, I recommend that you at least skim those more complicated sections. You will still be able to follow the flow of the book. And if you are interested in the subject in depth, I have supplied recommended reading in the endnotes.

Acknowledgments

I would like to thank the following individuals for help of various kinds in writing this book: Balázs Aczél, Kata Baka, Éva Bányai, Zoltán Baracskai, Judit Bokor, Zoltán Brandt, András Czellér, Péter Fábri, Zoltán Gazsi, Péter Gelléri, Dániel Gönci, Márta Hadházy, Katalin Kálmán, Balázs Karafiáth, Erika Kovács, Éva Kovácsházy, Sándor Kürti, Gábor Ligeti, Bálint Madlovics, Béla Marián, Kinga Massányi, Aposztolisz Mavromatisz, Csaba Mérő, Katalin Mérő, Vera Mérő, Anna Pavlov, Attila Pór, Eszter Rácz, András Simonovits, Tamás Sipos, Krisztina Somogyi, István Szamosközi, Gábor Szász, János Száz, Zsuzsanna Szvetelszky, Zsuzsanna Takács, Péter Tátray, András Telcs, Gábor Telcs, Katalin Varga, Tamás Varga, Zsadány Vécsey, and Jolán Velencei.

Special thanks should go to József Bencze, who prepared camera-ready figures. I wish also to thank the copyeditor (who, like "the number who lived faithfully a hidden life," prefers to remain anonymous) for a thoughtful and thorough job of copyediting, as well as the following individuals at Yale University Press: William Frucht, for his careful reading and meticulous editing of the manuscript; Karen Olson, for her assistance and support in preparing the manuscript for editing; and Ann-Marie Imbornoni for managing the editing and production of the book.

For preparing the English version, I thank Márton Moldován for his precise translation and David Kramer for what he has contributed: not only did he add literary flair to Márton's translation, but he also cleaned up many of my vague or imprecise locutions and greatly increased the book's accuracy, lucidity, intelligibility, and, what is most important, the degree of pleasure that I hope the reader will derive from this book.

PART ONE
Secular Miracles

The world may be one, but it has
endless ways of revealing itself.

1

On the Existence of Miracles

The Big Bang, luckily for us, is not a reproducible event. If it were,
we would all be dead.

My childhood friend Alex, who later became a successful businessman, doesn't believe in miracles. Nevertheless, one of his major
ambitions, which he considered a matter of patriotism toward the
small Central European country where we grew up, was to register a
Hungarian enterprise on the Nasdaq stock exchange. It didn't matter whether the company dealt in medical technology, energy supply
networks, or computer games. It just had to be new, original, and
marketable worldwide. If he saw any possibility in a firm, he was
willing to invest in it. And that was how he began to grow a little
business of mine that had been quietly struggling to survive.

Alex took over the management of my firm, and it was then that
I saw how amateurishly I had been doing things. He didn't, however,
interfere in technical matters, although I made severe mistakes in
that area as well. His attitude was the same as that of IBM founder
Thomas J. Watson, who was once approached by an executive con-
fessing to an error that had cost the company $10 million. "Go
ahead," said the executive. "Fire me. I deserve it." "Fire you?" Watson
responded. "I just spent ten million dollars educating you."[1]

When something went wrong, Alex would quote his business mentor, who was fond of saying, "There is always the next challenge." The chance that a miracle was waiting just around the corner kept him working with enormous enthusiasm and devotion. Of course, Alex wouldn't have called what he was waiting for a "miracle." I don't think a week has gone by in the last thirty years without his saying at least once, "There are no miracles." Perhaps that is why the miracles he sought so passionately never materialized. He never did get a Hungarian enterprise listed on the Nasdaq, although there is one now, called LogMeIn. Nevertheless, his efforts launched many enterprises, institutions, and investments that have proven successful and sustainable.

I maintain that miracles exist in our world; if so, it means that they should, in theory, be accessible to scientific study. But how are they to be studied if what makes them miraculous is exactly that they are one-time events and thus by their nature irreproducible, and unreachable by scientific analysis?

What is known as the scientific method relies on the formulation of models that can be verified through systematic observation, measurement, and experimentation under reproducible conditions. If the experiments confirm the model and the model produces miracles—that is, unexpected, extremely unlikely phenomena—then the analysis of such a model should provide a purely scientific way to study what we are calling miracles. For example, the Big Bang, luckily for us, is not a reproducible event—at least not by us. If it were, we would all be dead. Nevertheless, scientists are able to study the Big Bang with the aid of models that can be formulated, tested, and refined by reproducible experiments. We will take a similar course in this book, hoping, perhaps, to convince even Alex that there are in fact such things as miracles, even if we don't quite believe in them.

Black Swans

Nassim Nicholas Taleb, author of *The Black Swan* and other successful books, hails from Lebanon, a country that in my youth was known as the "Switzerland of the Middle East." For many years, Lebanon was an island of peace and prosperity in a region rife with blood and turmoil, and it remained so while the two great Arab-Israeli wars of 1967 and 1973 raged next door. In 1975, a riot broke out, but the Lebanese elite, including Taleb's family, failed to recognize its significance. They assumed that the riot would run its course and order would be restored in a few days. But order was not restored. Lebanon descended into civil war, and Taleb and his family eventually left the country. To this day, it is a source of wonder how a flourishing, prosperous country could have collapsed overnight.

Taleb reached America and studied at the Wharton School, one of the top institutions for business education, where he was taught the latest theories of modern finance. But there was much that he disagreed with. He believed what had happened to Lebanon could happen at any time to any complex system—including financial markets—and as an investor he had no intention of letting his portfolio end up like Lebanon. When he had made enough money to launch his own investment firm, he adopted an unorthodox strategy. He figured he could win big by betting that sooner or later, there would be a sudden collapse somewhere in the world economy, like the 1975 collapse in Lebanon. (In chapter 6, "The Sources of Equilibrium," I will discuss in detail the exchange structures on which Taleb's strategy was built, and I will give a detailed analysis of their advantages and disadvantages in chapter 10, "Life in Wildovia.")

Taleb used what he had learned at the Wharton School in a way that put him totally at odds with most investors. He wasn't interested in earning comfortable short-term profits by taking advantage of daily fluctuations in the markets. His goal was to reap enormous rewards should the economy suffer a major crash, while keeping his daily losses to a minimum until such an event occurred. He would not suddenly go bankrupt, he said, although he might slowly bleed to death if nothing cataclysmic happened for a long time.

Taleb's strategy is far from new. It has been viewed with suspicion for centuries. Various forms of short-selling—betting on the decline in value of a financial instrument—were first prohibited in the 1600s, and whether or to what extent short-selling should be permitted is a disputed question today. A number of countries instituted restrictions on short-selling during the financial crisis of 2008–2009.[2] The justification for such regulation is that a strategy of short-shelling can act as a self-fulfilling prophecy, itself contributing to a financial crisis. In chapter 11, "Adapting to Wildovia," we shall see that a total ban on short-selling can hurt the economy more than it helps, yet some regulation is reasonable. Regulation has received increasing emphasis in recent years, but Taleb still has plenty of scope for continuing his investment strategy.

On September 11, 2001, when global stock markets crashed in the wake of the attack on the World Trade Center, Taleb became a very wealthy man overnight. He continued to employ his investment strategy while seeking a better understanding of market crashes, and he continued to develop the software he had designed to implement his strategy. He could now afford to play for higher stakes, and he could afford greater short-term losses. The global financial crisis of 2008 made him a billionaire.

Taleb stated in 2008 that although he has made billions, there is much about which he remains as ignorant as before.[3] He still doesn't understand why Lebanon collapsed in 1975 or why the global economy did so in 2008. He knows only that one must be prepared for the occasional occurrence of unpredictable, even almost inconceivable, events. Taleb is now an academic researcher who studies catastrophic events—let us call them negative miracles—but his ideas apply to many positive miracles as well.

Taleb calls such negative events "Black Swans." *The Black Swan* came out in 2007 and topped the *New York Times* bestseller list for seventeen weeks. (I should note that the book has nothing to do with the like-named 2010 Darren Aronofsky film or the 1954 novella by Thomas Mann.) Despite the efforts of hardworking editors to make it more or less presentable, Taleb's book is a confusing, poorly written gallimaufry of ideas, thrusts, and parries. The secret of its success is that behind the confusing ramblings on financial philosophy and the flood of arrogance pouring out of the author's hyper-inflated ego (every Nobel laureate in economics is an ignorant quack, George Soros and all the other Wall Street hotshots owe their success to sheer luck, and so on) is a man who, one cannot help but feel, knows something important about the world. His execrable style aside, Taleb is genuinely smart, and he knows his stuff, including the mathematics.[4]

Taleb uses the idea of a black swan to represent things that, though almost inconceivable, nevertheless occur and have a significant impact on the world. The black swan is an unfortunate symbol, for it is something that we can easily imagine, even if we never saw one and had no idea such a thing exists, since we consider it part of a swan's essence to be white. Nonetheless, we have no difficulty picturing a black swan: it is just like a normal swan, only black. And why not? Indeed, there is a species of black swans living in Australia, but that is beside the point. The book's German publisher didn't even put a black swan on the cover. The cover of *Der Schwarze Schwan* boasts an origami swan—in hot pink.

Nowhere in more than five hundred pages does Taleb define the term "black swan." Instead, we are given numerous examples of what he considers "black swans," including the collapse of Lebanon, the dissolution of the Soviet Union, the fall of the Berlin Wall, the birth of Islam, and great stock market crashes. But there are also smaller-scale examples. The inexplicable success of a book can be a black swan, as can an unexpected flop. Other black swans include the many lovers of Catherine the Great (Taleb tells us in a footnote that the precise figure is twelve, not that many by today's standards); the 2004 tsunami in the Indian Ocean; the invention of the computer, the laser, the Internet, and even the wheel; the discovery of America and that of antibiotics. To Taleb, anything that is unpredictable via common sense and science, but that occurs nonetheless and has a great impact on the world, can be a black swan.

In this book, I will use the term "miracle" in a sense similar to Taleb's use of "black swans." Like Taleb, I will not give a precise definition, so suffice it to say that a miracle's most important characteristic is that it is a one-time, unpredictable and irreproducible event. Our miracles need not be as grandiose as Taleb's black swans: they do not necessarily change the world. There are major miracles and minor ones, but all of them are miracles nonetheless.

Miracles and Black Swans

Just because it is a one-time and irreproducible occurrence doesn't mean that a miracle cannot occur again. It may reoccur, as mysterious and irreproducible as it was the first time.

I am inclined to think that every person is actually a minor miracle, being a one-time and unrepeatable occurrence. I am also inclined to agree with the scholar Martinus Biberach, known for having copied out a well-known quatrain in 1498, but also with Martin Luther, who dubbed the poem the "rhyme of the atheists" and revised it to be more theologically correct. Here is the original poem:

I live and know not how long,
I die and know not when,
I travel and know not whither,
I marvel that I am happy.

And Luther's riposte:

I live as long as God wills,
I die when and how God wills,
I travel and know precisely whither,
I marvel that I am sad.[5]

Biberach's happiness and Luther's sadness can both be considered minor marvels—or miracles—though my own worldview puts me closer to Biberach.

A universal characteristic of black swans is that after one has occurred and become known, it may in retrospect seem inevitable. We generally hear many explanations after the fact. Although black swans and miracles are similar in many ways, they differ in this respect: a miracle frequently defies all logical explanation, even in hindsight.

For the faithful, miracles in the strong, theological sense of the word serve as evidence of the power of God, so the believer has no reason to doubt their existence. If something previously considered a miracle (such as solar eclipses) is found by science to be a consequence of the laws of nature, the believer's faith is unshaken. Man is fallible and may mistake one of God's laws for one of God's miracles. Nevertheless, there are many miracles that defy explanation, such as the virgin birth, and one must, in the words of Tennyson, "by faith and faith alone embrace, believing where we cannot prove."[6]

For the nonbeliever, what we call miracles only demonstrate the limits of our knowledge. There have always been phenomena that our current knowledge cannot explain. But that doesn't make them theological miracles. Sooner or later, science will find an explanation. But science has also proved, while providing explanations for more and more natural phenomena that previously defied understanding, that there will always be phenomena that science, however much it advances, cannot explain. I will discuss this in a later chapter, where I will also define more precisely what we mean by miracles.

Essential Reading

Douglas R. Hofstadter's *Gödel, Escher, Bach* ranks among the books that have had the greatest effect on my thinking, even though I thoroughly reject the author's premise. The book argues, over seven hundred–plus pages, that the system of logic discovered by the Austrian logician Kurt Gödel (which I will describe in detail in chapter 3, "The Source of Miracles: Gödel's Idea") is a sufficient basis for creating a full-fledged artificial intelligence. Moreover, Hofstadter had the chutzpah to write in his introduction, "In a way, this book is a statement of my religion."[7] Let's just say that I attend a different house of worship. Nevertheless, most of what I know about Gödel's theorem and logic in general comes from Hofstadter, for it was he, unlike the hordes of experts whose articles he cites, who told the story of Gödel's theorem in a way that captivated me. Hofstadter revealed for me the source of miracles, and I could not have written this book without having read his.

Everything you need to know about Gödel's theorem can be obtained from Hofstadter's book. You don't have to reinvent the wheel when it comes to understanding scientific facts that have been established beyond doubt. You simply have to learn them, and from books like Hofstadter's if possible. The Bible is also a primary source for the way I think about the world, even though I am certain that the birds and the beasts were not created as described in Genesis. We simply cannot think about flora and fauna as we did before Darwin, just as we cannot think about the Sun and Moon as we did before Copernicus. In fact, every one of the books that has shaped my worldview—including fictional works by such writers as Géza Ottlik and J. K. Rowling—is a work that I disagree with in some fundamental way. But I still take great pleasure in letting these stories shape my thinking.

The same goes for *The Black Swan*. I fundamentally disagree with Taleb's point of view. Possibly it is just my prejudice, because my home country, Hungary, has never collapsed overnight, at least not in the past four or five centuries. Then again, it has also never been the Switzerland of Central Europe. I don't believe that the world is as wild and fragile as Taleb thinks, and that is why this book devotes about as many words to the mild world as to the wild world. We shall see that black swans occur in both places.

I do not recommend trying to learn about black swans from Taleb. His book is just too messy and confusing. But it provides an endless trove of witty insights and trenchant examples, and it will likely change the way we think about the world for years to come.

The Miracle of Rubik's Cube
In 1971, Ernő Rubik became a professor of architecture at the Hungarian University of Art and Design in Budapest. He found

himself continually surprised and dismayed at his students' poor spatial imagination. In thinking about how to improve the situation, he decided to develop some sort of device that would help his students improve their capacity for three-dimensional visualization. After a couple of years' thought and experimentation, he came up with Rubik's Cube. Hundreds of millions of them have been sold worldwide.

From 1974 through 1979, Rubik sold five thousand of his cubes in Hungary but only another two thousand in the entire rest of the world. It was the unanimous opinion of game industry experts that there was no world market for his invention, even though it had captured the imaginations of thousands of Hungarians. No matter how hard he tried, Rubik could not make a dent in this opinion.

I never was able to solve the puzzle on my own, but like most of those bitten by the Rubik bug, I learned a few patterns of rotation by word of mouth, and then I was able to enjoy twisting out the solution again and again. My delight was never marred by the fact that I invariably needed four or five minutes to solve the cube, whereas speedcubers can do it in under ten seconds. Most of them know thousands of combinations of rotations rather than the three I had at my command.

At a time when the cube was known to only a few initiates, I asked a friend, a first-rate mechanical engineer, whether one could design a mechanism that could be twisted in all three dimensions. He replied with complete certainty that such an object could not exist. I pulled out my cube. He twisted it for a while, pondered a bit, twisted it some more, and then theatrically held it aloft and proclaimed, "Gentlemen, this object does not exist!"

In 1979, after many unsuccessful attempts, Rubik showed the cube to Tom Kremer, holder of hundreds of game patents and head of a game developing and trading firm in London. Tom twisted it for a couple of minutes, then, holding it up the same way the mechanical engineer had done, proclaimed, "This puzzle contradicts every principle known to the industry. It doesn't make any noise, it doesn't look valuable, it isn't cute, and no normal person can solve it." This pretty much summed up why it had been so unsuccessful on the world market. Then he continued: "This thing is ingenious. Let's do it fifty-fifty." Thus did Rubik's Cube begin its march toward becoming a black swan.

Rubik's Cube isn't the only miracle in this story. Tom Kremer's keen eye is another. He was just a year short of fifty in 1979 and already a veteran of the game industry. Thirty years later, he said that discovering the genius of Rubik's Cube and making it world famous was the high point of his professional life. These two miracles—the cube and the keen eye, neither one by itself a true black swan—together formed a full-fledged swan.

Rubik's Cube was a miracle even before it was apparent that it would become a black swan. It was a miracle to those whom it had captivated. The mechanical solution was a miracle to engineers. And Tom Kremer, veteran of the game industry, could also see that it was a miracle.

The main difference between black swans and miracles is this: like a black swan, a miracle is a one-time, unrepeatable occurrence, but it doesn't necessarily have a huge impact on the world. There are minor miracles that have little effect on the world, and there are major, world-shattering miracles.

Vocation

Ernő Rubik realized his dream of constructing a device that could be freely twisted in three dimensions. That would have been a miracle in itself even if there had been no Tom Kremer, and the cube hadn't become a black swan as an unexpected worldwide success.

The English sailor Ellen MacArthur was also someone who realized her dream. In 2001, at the young age of twenty-four, she completed a solo circumnavigation of the globe. When she arrived home after ninety-four days at sea, she said that she hoped her example would encourage other young people to fulfill their dreams. Her feat was not unique; she actually came in second in a solo round-the-world race, the Vendée Globe. Three years later, however, in a boat specially designed for her, MacArthur set a world record for the fastest solo circumnavigation. (The record stood for only a few years, but who cares about that? It doesn't diminish her achievement.) Thus did a miracle manifest itself through the dreams of a young Englishwoman, just as it did through Ernő Rubik's endless experimenting. The actual form that a miracle takes is of little importance, whether it be a young woman's solo circumnavigation of the globe, the creation of a theoretically impossible object, or even the parting of the Red Sea. These events were all unimaginable before the fact, and when they occurred, they caused a sensation. We shall therefore consider them miracles.

The great Hungarian prose stylist Géza Ottlik (1912–1990) writes the following in his novel *School at the Frontier:*

> That's how it always is: nothing goes right, hundreds and thousands of one's wishes and hopes come to nothing; but there are one or at most two important things, without which life couldn't go on, which always work out in the end. In a casual, accidental sort of way, of course. Destiny can dispense with our gratitude.

Later, I stopped getting excited about this sort of thing; I knew it was quite superfluous to worry about essential things—for instance, that I should become a painter; I knew I'd do that even if I had to change the position of the Milky Way for it; I knew I'd walk comfortably through the thickest stone wall and that the Red Sea would divide for me.[8]

A big dream and stubborn persistence can be enough for a miracle to emerge—and "destiny can dispense with our gratitude." But what should you do you if you lack such ambitious dreams? The thoughtless answer is that you should find an ambition and set out to realize it. That is a surefire way to make many, many people unhappy who were not destined to be unhappy. Not everyone is granted the gift of a dream, and of those who have it, only some can make their dreams come true. You can lead a happy, purposeful, and satisfying life without miracles.

I have always envied people who decide at age ten that they are going to be molecular biologists, airline pilots, sailors, or painters, and then manage to do it. For me, a career in mathematics seemed to be a given. I was successful as a boy in mathematical competitions, and I went on to become a professional mathematician. For a while, I dreamt of proving one of the great unsolved conjectures: the four-color conjecture, the Poincaré conjecture, the Riemann hypothesis, or Fermat's last theorem. What a miracle that would have been! But I quickly realized that there were many mathematicians much more talented than I, and if they couldn't solve those famous problems, it was hardly likely that I would.

A miracle occurred in three of the four cases, though not for me: in the forty years since I took my degree, three of those problems have been solved. But hundreds of other mathematical problems—just as difficult though perhaps not as famous—continue to parry every thrust at their solution. You don't read about those problems in the newspaper, nor do you hear about the thousands of talented mathematicians who may have spent entire careers trying to solve one. Some of them may proudly, and justifiably, assert that although they themselves didn't succeed, they contributed, even if only a little, to someone else's eventual success. But the majority can't even claim that much, for the approach they took to solving the problem proved to be a dead end.

Such is the human condition. Many must follow a will-o'-the-wisp into a blind alley so that a few may fulfill their dreams. For every miracle there are countless failures and disappointments. But though your only contribution to others' success may be to have explored a few promising dead ends, you may still have done a good life's work. You will have led a productive—though miracle-free—life, done a decent job, and achieved, one may hope, a measure of happiness and contentment.

Full-bore commitment to a big dream may not be strictly necessary to create something important. These days, miraculous and important scientific results and technological advancements most often emerge from the teamwork of hundreds, even thousands, of individuals. A team will quickly disintegrate if every member dreams too big. A few visionaries are necessary if miracles are to be achieved, but so are armies of seasoned professionals with expertise in some field who can solve well-formulated problems.

There is an article by two Australian professors, Gerry Mullins and Margaret Kiley, with the self-explanatory title "It's a PhD, Not a Nobel Prize."[9] I require all of my doctoral students to read it, because it helps many of them moderate the lofty aspirations that prevent them from writing a competent dissertation and moving on to become highly qualified contributors to their professions.

Variants of the following sentiment have been attributed to a number of famous people: if you are not a communist (or socialist or democrat) at twenty, you have no heart, but if you are still one at thirty, you have no brain. It is natural and appropriate to have big dreams and noble ideals in our youth, but sooner or later we must temper those dreams with the knowledge of the limitations of our own abilities. Ellen MacArthur's way—fulfilling a lifelong dream— is just one option. If we are not one of the very few who have a chance of fulfilling our big dream, then we must find a more realistic garden to cultivate.

MacArthur's endeavor failed to become a black swan because it did not have a particularly large impact on the world. Circumnavigating the globe has not by itself been a black swan since Ferdinand Magellan's expedition in the first quarter of the sixteenth century, although it still seems like a miracle when it is done solo in such a small craft as Ellen MacArthur's. But she was not even the first person, or first woman, to do it. Had her success inspired many people to fulfill their dreams, then perhaps her feat would have become a genuine black swan. But other recent examples of great dreams fulfilled have employed thousands of qualified professionals—such as the dreams of the founders of Microsoft, Apple, and Google.

Harry Potter's Miracles

In another of Ottlik's stories, the prince has slain the dragon, built the castle, kissed the princess to awaken her, and made her laugh. And of course they were married at once. But Ottlik's fairy tale continues:

But a day passed, and another one, and another. The first day was long enough already; it consisted of many hours, the hours of many minutes, the minutes of many seconds.

They looked at each other, and the prince decided to wall in the princess, put her back to sleep, demolish the castle, and revive the dragon.

He began right away. But it wasn't as simple as that. Neither hard work nor even more heroic deeds, neither adventures nor running about helped him destroy the castle, which kept turning away from him on its duck's legs. Even less was he able to restore the dragon's severed heads, nor could he put the princess back to sleep or wall her in. Days passed, turning into months, and into years, and thus they struggled. And if they haven't died, they are still alive.[10]

Is that what happens when we realize our personal miracles? Do we find ourselves asking, "Is that all there is?" The reason I believe Harry Potter to be one of the great literary achievements of the past century is that J. K. Rowling performed the miracle of taking the next step.

Harry Potter is no Ellen MacArthur. He never had big dreams. He just happens to be the one to have performed a miracle that no one else could have performed and for which he himself was an unlikely prospect. And when, after seven volumes of adventure, he finally defeats the evil Voldemort, he lives on as a contented and useful wizard, with no further dreams, just as he would have done had he not been the only one who stood a chance of defeating him-who-must-not-be-named. He's a wizard, and miracles are his stock-in-trade, but he performs no further true miracles and is content with that. He carries on with his job, he loves his wife and children, and at times, he reminisces about some of the more adventurous episodes of his life.

Rowling performed another miracle with Harry Potter. she got millions of children to read thirty-five hundred pages in an age when we had almost given up hope that the younger generation might pick up a book rather than browse the Web and watch television. That mania for reading was just as sudden and unexpected as the fall of the Berlin Wall or a major economic crisis, and it may be as unrepeatable—that is, it may be a true miracle in our sense of the word. The difficulty Rowling had in finding a publisher also points in this direction. Just as Rubik's Cube was turned down by many toy manufacturers and distributors, Harry Potter was turned down by several dozen publishers, who have surely been kicking themselves ever since.

Nowadays, the success of a large project requires the contributions of skilled individuals who can solve problems in a professional way—the sort of person Harry Potter became when he grew up. Perhaps that is why Harry Potter's tale could have been written only in our time. Amid all those great adventures, I think Rowling knew from the first that this was what her books were about—that and, of course, the miracle of love.

We may ask what place is left for miracles as we see them being squeezed out by modern science, which is able to explain more and more phenomena that were once beyond human capacity to explain. Perhaps science will one day give us a complete description of Ernő Rubik's thought process, Ellen MacArthur's daring, or Harry Potter's remarkable and unexpected success, but it certainly cannot do so today. Can it nevertheless offer a way of analyzing the inexplicable, those one-time, unrepeatable events we are calling miracles?

2

The Mild World and the Wild World

A single Einstein has no standard deviation.

Albert Einstein bequeathed his brain to researchers after his death, but studies of its anatomy have revealed little of interest.[1] While Einstein's brain differed in many ways from any previously examined brain, it was impossible to discern the specific characteristics that combined to endow it with genius. To do so would have required studying the brains of a large number of Einsteins so that researchers could characterize a typical Einsteinian brain and discover how it differed from a typical non-Einsteinian one. But there was only one Einstein, and his brain turned out to be exactly as unique as everybody else's.

We generally think that "average" means "typical" and that deviations from the average are worth mentioning or investigating. But that is often not true. By any given measure—volume, weight, number and depth of convolutions, or what have you—our brains vary from the average just as our heights do. A man of average height in Hungary is five feet, nine inches tall, but there are few men of that exact height. Almost everyone is taller or shorter.

Because of these deviations from the average, statisticians, in addition to specifying the average value of whatever they are examining, also establish an average deviation from it. The statisticians' term for this is "standard deviation." The height of an average shorter-than-average man, for example—one standard deviation shorter than the mean—is five feet, six inches, and the average taller-than-average man measures six feet. In this sense, a five-foot-six man is just as "average" as a six-footer; each is one standard deviation from the mean. (Technically, what I am describing here is not exactly the standard deviation; mathematicians find it convenient to use a slightly more complicated formula.)[2]

We know from psychological studies that our language describes how we perceive things. We do not consider a man who is six feet tall remarkably tall, nor one who stands five foot six remarkably short. What most of us consider remarkable begins somewhere around twice the standard deviation from the average. Thus we regard six feet, three inches as tall, and when a man whose height is three standard deviations from the average—six feet, six inches—rises from a chair, it will surely command our attention. But this is still not a miracle. A sixteen-foot man, in contrast, would be viewed as a miraculous occurrence. His height would be forty standard deviations from the average. In fact, no one is sixteen feet tall—or even ten feet.

If someone talks about an average without mentioning the standard deviation, we should always be a bit suspicious.[3] Such a person is not necessarily trying to pull the wool over our eyes; he or she may be speaking out of ignorance. But keep in mind that an average without its standard deviation tells us little. Take, for example, the fact that the average child begins to speak at the age of eighteen months. Should we be worried about our little Monica's cognitive development if she is not yet talking at age two? It depends on the standard deviation. If the standard deviation for beginning to speak is a couple of months, then we should be concerned. In fact it is around six months, so Monica is an average late talker. When my mother took me to the doctor for this reason, he just said, "Have no fears, my good woman; he'll catch up soon enough."

The Genius as a Miracle

A single Einstein has no standard deviation. He is a unique event, impossible to explain through the methods of statistics. This uniqueness is also what makes the concept of genius so hard to pin down. Many people would say that a person with an extraordinary talent is a genius, but talent is a different concept.

Fortunately, there are many people with talent, so this concept can be examined using statistical methods. I like the following definition: to have talent is to know things or be able to do things that you never learned. Psychological studies with talented people have arrived at much the same notion, possibly phrased more professionally. Whether the subjects are high school valedictorians or International Science Olympiad gold medalists, the psychological profiles of talented individuals are essentially the same: the more you know beyond what you have learned, the greater your talent.

There are enormous differences in efficiency between people of different degrees of talent. An extraordinarily talented computer programmer, for example, can be more efficient than an averagely talented one by up to an order of a magnitude. That is somewhat surprising. It is as if a man who is six feet, three inches were found to be ten times more efficient—or ten times more anything—than a man who is only six feet tall. It appears that height doesn't have such radical implications, even in basketball, where it is a positive asset. Talent, however, makes a huge difference.

What should we do, then, with the idea of genius? We could perhaps use it to mean an extraordinarily high degree of talent. When
people talk about a "genius-level IQ," they are defining it on a linear
scale: genius is like ordinary intelligence, only a lot more so. But perhaps it is something qualitatively different from either intelligence
or talent. Or are we better off defining a genius as someone who
conceives of things even the most extraordinarily talented person
would never dream of? Think of Newton or Einstein, Mozart or
Picasso.

While the notion of a spherical Earth existed in antiquity and
the idea of gravitation as a measurable force goes back at least to
Galileo in the late sixteenth century, it was Newton who made it possible to understand how a round Earth might work, why the people
on the "bottom" don't fall off—or if they happened to stick to the
Earth's surface somehow, how they could live upside down without
all the blood rushing to their heads. It was indeed a puzzle how people could live happily upside down. The church was not being completely pigheaded in not accepting a spherical Earth that revolved
around the Sun. Not only did such ideas go against its theology, but
also there was really no good answer to the scientific questions that
such a cosmology raised. Newton solved the problem by coming up
with the law of universal gravitation: the idea that the same force
that kept the planets revolving around the Sun also kept people's feet
on the ground. For that we needed a genius, someone to discover a
concept no one else had come close to imagining and to give a unified explanation for phenomena that had previously seemed both
unrelated and inexplicable.

In this sense, genius seems like something unique and unrepeatable—by our definition, a miracle. And so we see again that miracles

indeed exist, because every now and then, but not too often, a New-ton or an Einstein is born. These are individuals who are more than extraordinarily talented. Not only do they know much more than they ever learned, but they also know something that cannot be taught, because it is beyond their contemporaries' wildest imaginations.

An ingenious new idea today will become tomorrow's elementary knowledge. A theory that only a genius could have invented will someday be taught by thousands of hardly ingenious teachers. At first, when the ideas are still new and revolutionary, it takes a great teacher to understand and pass on the new knowledge. But over time, those interested in the theory refine and simplify it and discover new ways of explaining it that make it much easier to teach. Newton's original *Principia* is very difficult to comprehend. I have read it, and I can testify that it is unreadable. But the material in it is taught every year to thousands and thousands of high school seniors and college freshmen. A genius breaks new ground, and extraordinarily talented people follow the trail the genius has blazed. Later, moderately talented people follow in their footsteps.

I have been arguing that a genius is a miracle, yet I still have no answer for my friend Alex, who is convinced that there are no such things as miracles. Alex would say, "A miracle to you is not necessarily a miracle to someone much more talented than you." And sure enough, a five-meter long jump qualifies as a miracle for me, yet there are athletes who can jump over eight meters, so even an eight-meter long jump is not considered a miracle.

It will be helpful to try to quantify somehow the rarity with which miracles occur. So let us visit two imaginary countries where we can learn about mild averages and wild extremes. Let us call them Mildovia and Wildovia, the mild world and the wild world.

Mildovia and Wildovia

Try to guess the answers to the following questions:

What is the average height of someone taller than six feet seven inches?

What is the average age of someone over ninety?

What is the average net worth of a "very high net worth" individual, someone with at least five million dollars in liquid financial assets? (There are perhaps twenty million such people in the world.)

What is the average capital stock valuation of a company that holds over $5 billion in assets? (There have been seven to eight hundred of these worldwide in recent years.)

The first and the last of the above questions appeared in Taleb's *The Black Swan.* The answers, however, did not. We are not going to be so stingy: the answers are six feet eight inches, ninety-three years, $80 million, and $27 billion.

What we begin to see from these answers is that the mathematics involved in the first pair of questions is radically different from what is involved in the second pair. They seem to come from different worlds, even though both pairs of questions concern concrete pieces of numeric data. So how do they differ? The first thing we notice is that there are no really large deviations from the average in the case of height and old age. No one is five yards tall, and there is no one who, like the biblical patriarch Methuselah, lived to be 969. On the other hand, there are most certainly some very wealthy individuals, such as the notorious Koch brothers, each with a net worth of around $40 billion. And there are extremely valuable corporations, such as Apple, which attained a market valuation of around $750 billion in

August 2012. The tides of business were such that Apple sank below $450 billion three months later. Such a percentage loss would have been devastating for most individuals, but Apple survived and recovered easily.

Taleb calls the country where no such large deviations from the average occur "Mediocristan," and the place where monstrous anomalies are to be found, "Extremistan." I have nothing against the name Extremistan, but I find "Mediocristan" unfortunate. Taleb looks on Mediocristan with disdain. He rejects its methods and its ways of thought, even though they produce successful outcomes. It appears that his experience of seeing Lebanon turn suddenly, after more than a century of peace and prosperity as a Mediocristan, into an Extremistan of violence has taught him to hate all that is average and to doubt the theories of such a place.

I prefer to give the place that is free from wild deviations the more neutral name "Mildovia." And if we are going to have a Mildovia, let us rename Extremistan as well: "Wildovia." Unlike Taleb, I do not believe that the science of Mildovia is mere smoke and mirrors, and I continue to teach Mildovian methods proudly, keeping in mind that they apply only to some parts of the world and that where they fail to apply, they tend to lead us astray.

We shouldn't be talking about Mildovian science and Wildovian science as if they were different things. There is one scientific method, one mathematics. It is just that the two types of phenomena are described using different mathematical models. And there is plenty of charlatanry about both, including the showcasing of spectacular averages without any mention of the standard deviation.

We take it as a matter of course when biologists use one model to describe the behavior of wolves and another for the behavior of sheep. One model concerns a predator in the wild, the other a grazing farm animal. Each model has its domain of validity, and it does not surprise us that a model that satisfactorily describes the behavior of wolves fails when applied to sheep.

Even though biologists might require different models for wolves and sheep, and physicists might have one model for the interactions of subatomic particles and another for the movements of the planets, all scientists adhere to the scientific method: we observe, we experiment, we set up models based on experimental results, and we try to establish the domain of validity of each model. A successful model allows us to predict events within its domain of validity. Sometimes a model devised to describe one set of phenomena can be applied to another domain entirely, such as when the field of economics successfully adapted models from physics.

Life is slow and mellow in Mildovia, with no big surprises. There are no large deviations from the average. Although Mildovia has tall people and short people, smart people and not-so-smart, it boasts no five-yard giants, no superrich masters of the universe, no geniuses. Yet there is much going on in that mild world, and the efforts of many talented thinkers were necessary to achieve an understanding of the broad range of phenomena that occur there. Our understanding of Mildovia forms the basis for most of our understanding of the universe, and we should not discard our models of the mild world just because there exist phenomena they cannot adequately describe, such as the behavior of the world economy in the last few decades.

Across the border in Wildovia, things can get hairy. There you can find enormous deviations from the average, such as Apple, which even with its diminished capital topped the 2013 list of corporations, with more than twice the capitalization of runners-up Shell and IBM. In Wildovia, values that are forty standard deviations from the mean are commonplace. Apple's valuation is more than four hundred times that of an average corporate giant.

Experienced pub-goers can tell when a fight is about to break out. The smooth flow of conversation is suddenly broken by angry words. At that point, the balance tips, and the peaceful world of mild-mannered patrons keeping to themselves turns into a brawl. Something similar happens in science. You can tell that an accepted scientific model is about to be challenged when more and more phenomena are discovered that the model should be able to describe but cannot, and baffled scientists begin to raise their voices. It is out of this bafflement that radically new models emerge.

The Normal Distribution

The German mathematician Carl Friedrich Gauss (1777–1855) was called the "prince of mathematicians" by his contemporaries. One of his important inventions was what is called the normal distribution, also called the Gaussian curve or, less accurately, the bell curve (see figure 1). The normal distribution proved to be essential in describing Mildovian phenomena. Phenomena that are normally distributed have most of their values close to the average, and the farther you go from the average, the scarcer the values become. For example, if the normal curve in figure 2 represents the heights of Hungarian men, we would expect to find that about two-thirds (68 percent) fall within one standard deviation of the mean height of five feet, nine inches—that is, between five feet six inches and six feet—while fewer than one-tenth of one percent would fall outside three standard deviations from the mean—that is, taller than six feet, six inches or shorter than five feet.

The Greek letter mu (μ) labeling the midpoint of the horizontal axis stands for the average, or to use the more precise term, the *mean.*

Figure 1. The last German ten-mark note (before the mark was replaced with the euro) with a portrait of Gauss; pictured also is the Gaussian normal curve.

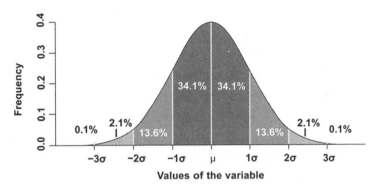

Figure 2. The Gaussian curve, or normal distribution
(drawing by József Bencze).

As you can see, the curve attains its maximum value at μ, which means that for the normal distribution, the mean value is also the one that occurs most frequently. The Greek letter sigma (σ) denotes standard deviation. You can also see that 34.1 percent of the population has its values of whatever is being measured (height, for example) between the mean and the mean-plus-one-standard-deviation.

(Another 34.1 percent are one standard deviation or less below the mean.) Also, less than two-tenths of one percent of the population (one in five hundred) falls three standard deviations away from the mean. That is how values are distributed under the Gaussian curve. I will spend some time in the second part of the book praising its descriptive power. For now, suffice it to say that this distribution models many natural phenomena very well.

My friend Alex was right about miracles insofar as they relate to normally distributed phenomena. The normal curve falls off so rapidly that by four times the standard deviation, its value is so close to zero that you would need a powerful microscope to see the space between the curve and the horizontal axis. After ten standard deviations, not even a microscope would help. Only one in a trillion trillion instances is expected to exceed the mean by ten standard deviations or more.

Because the Gaussian curve had proved so effective in describing so many natural phenomena, it seemed reasonable to apply it to economics as well. Eventually, the statistical ideas behind the Gaussian distribution became such a dogma that for around a century, it never occurred to the creators of economic models to look anywhere else. But it turns out that the Gaussian distribution does not capture in full the way the economy works. And that is true not just of economics: many phenomena turn out to be outside the domain of validity of this particular model. During the 2008 financial crisis, I heard from various financial gurus that "A crisis like this could not have been expected even once in ten thousand years." Now, I am nowhere near ten thousand years old, yet I have heard that comment at least four or five times—for example, during the 1987 and 1998 crises and also after 9/11. Something must be wrong.

What is wrong is that the Gaussian distribution does not allow for shocks that seem to come out of nowhere. It predicts Mildovia

quite well, but it cannot deal with the tumultuous world of Wildovia. To describe such crises, we need a radically different model. In
fact there are such models, and they have been around just as long as
the Gaussian distribution, which provided such a solid basis for the
science of Mildovia.

Returning to our analogy of a fight breaking out in a pub, we
might say that Mildovians understand that when voices are raised
with increasing frequency, it means the peaceful world of Mildovia
is about to give way to the chaotic world of Wildovia. Whether we
are talking about barroom brawls or economic indices almost doesn't
matter. We are now within the domain of validity of the new model
we have created to describe the extreme situations of Wildovia.

The science of Mildovia became so sophisticated that it was
able to devise models whose scope goes beyond the usual Mildovian phenomena to describe the chaotic events of Wildovia. That
is one reason why we shouldn't look down our noses at Mildovian
science: the models of Wildovia were created by the science of Mildovia. The methods are precisely the same. It is only the models that
differ.

The Cauchy Distribution

The mathematician Augustin Louis Cauchy (1789–1857) does not
have his face on any banknotes, though he did appear on a French
postage stamp honoring his two hundredth birthday, and his name
can be found among those of the seventy-two great French scientists,
engineers, and mathematicians engraved on the Eiffel Tower. Like
Gauss, Cauchy is among the names most frequently encountered by
students of engineering and mathematics. Some 150 years after Newton, he cleared up many of the ambiguities in differential and integral calculus and brought the subject into a form that can be taught
as an introductory college course.

The curve constructed by Cauchy (figure 3) can be seen in figure 4, and at first glance, it looks a lot like the Gaussian curve. (When I noted above that "bell curve" was not an accurate name for the Gaussian distribution, I had in mind the Cauchy distribution and many other curves that are also bell-shaped.) Nothing seems to suggest why it should describe a much wilder world than the Gaussian curve.

A closer inspection shows that Cauchy's curve is not as close to the horizontal axis at three standard deviations as Gauss's curve, though it does get closer and closer for larger and larger values of the standard deviation. Does it really matter how fast the curve approaches the horizontal axis? Can the very nature of the model that a curve describes depend fundamentally on the rate at which it approaches zero? We will soon see that it can.

Figure 3. Augustin Louis Cauchy (1789–1857), French mathematician and physicist.

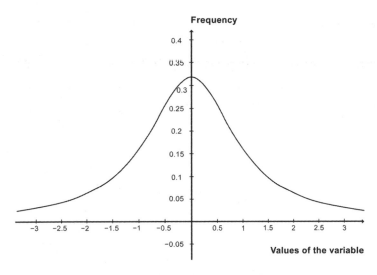

Figure 4. The Cauchy distribution (drawing by József Bencze).

Of all the mathematical and physical phenomena that lead to a Cauchy curve, the following is perhaps the easiest to understand. Suppose a woman with a rifle—let us call her Phoebe, in honor of the great American markswoman Phoebe Ann Mosey, better known as Annie Oakley—is standing at some distance—let us say ten meters—from a wall that stretches to infinity in both directions. She closes her eyes and spins around, and when she stops, at some random angle to the wall, she fires her rifle in the direction in which she happens to be facing. Half the time, of course, she won't hit the wall at all, since she will be facing away from it, but let us just look at the shots that do strike it. Most of the time, Phoebe's bullets will hit the wall relatively close to her. Half of her shots will be within 45 degrees of perpendicular in either direction and will hit the twenty-meter section that is closest to where she's standing. So like Gauss's curve, the tallest part of Cauchy's curve is in the middle. But if Phoebe happens to spin herself almost parallel to the wall, then her bullet will strike a much more distant point. The Cauchy distribution describes how often, on average, each point of the wall will be hit.[4]

The main difference between the Gauss and Cauchy models is that in the Gaussian distribution, the very distant parts of the wall are extremely safe. If Phoebe stands ten meters from the wall and fires once a second with her angle of shooting governed by the normal distribution, it will take her 10,000 years on average for a shot to reach 65 meters or more, while with the Cauchy distribution, it will take her on average only 21 seconds to reach 65 meters. Furthermore, it will take only 5 minutes on average to reach 1,000 meters or more, 52 minutes for 10,000 meters, 9 hours to reach 100 kilometers, and only 3.6 days to reach 1,000 kilometers. Thus in the Gaussian case, we are safe at a distance of less than 100 meters from Phoebe's gun, while in the Cauchy scenario, we are not safe even a thousand kilometers away. Rarely, but within a reasonable time, every region of the wall will be hit.

Chris Anderson's well-known book *The Long Tail* argues that in today's economy, business opportunities are to be found precisely in the areas far from the mean. Anderson proposes a strategy of finding a sufficiently large distribution channel in which one may market a large number of unpopular items rather than a small number of popular items. The Cauchy curve provides a model for such a strategy. In contrast to the Gaussian curve, opportunities far from the mean are by no means rare. Figure 5 shows both curves together, and if you look, you can see why Chris Anderson chose his title. The fat part of the tail of the Cauchy curve is much longer than the fat part of the tail of the Gaussian curve. It is not that one tail is longer—both stretch out to infinity—but the Gaussian curve quickly becomes so thin that for all practical purposes it disappears.

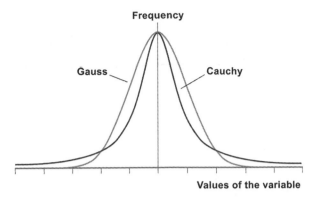

Figure 5. The Gaussian and Cauchy distributions (drawing by József Bencze).

When Cauchy set out to investigate his curve's mathematical properties, he began by investigating the mean value of the location of Phoebe's bullets. The answer seemed obvious enough: it must be at the center of the wall since it is equally probable that Phoebe will stop spinning facing to the left as to the right. And indeed the curve is symmetric, but when Cauchy tried to calculate the mean expected value following a finite number of shots—say ten, a hundred, or a thousand—he found that as the number of shots increased, the more probable it became that for one of those shots, Phoebe will have faced almost parallel to the wall, thereby hitting an extremely distant point, and that point would not be canceled out by the remaining shots. Thus the average position of many shots fails to approach the middle of the wall; rather, it jumps all over the place, under the strong influence of an extremely distant shot.[5]

As a mathematician would phrase it, the Cauchy distribution has no expected value. The average of a large number of shots can be at any point along the wall. This is precisely what does not happen with a Gaussian curve. As more shots are fired, the mean value of the

Gaussian curve comes closer and closer to the middle of the wall as the very rare distant shots are ultimately canceled by the much larger number of shots close to the center.

The Cauchy distribution also has no standard deviation. And the way it fails to have one is different from the way a single Einstein doesn't have one. It would be more accurate to say that Einstein does, in fact, have a standard distribution, except that it is equal to zero, and therefore it has no reasonable interpretation. The Cauchy distribution has no standard deviation in the sense that no number is large enough to measure the typical deviation of the shots from the center. In everything that is well modeled by the Cauchy distribution, it is not only a unique, unrepeatable Einstein that has no standard deviation; it turns out that the entire population has none.

Cauchy's concept led to a wild world in which we cannot even talk about such "obvious" things as the typical deviation from the typical value—because there are no such things: no typical value (the mean) and no typical deviation. There is no way of saying what is "average." If you want average, you have to leave Cauchy's world behind and return to the safe and mild Gaussian world.

We may say, then, that the mathematics of Mildovia comes from Gauss, while the mathematics of Wildovia comes from Cauchy. What is rare in Gaussian Mildovia is relatively common in Cauchy's Wildovia—for example, a shot landing much, much farther away than the eye can see. If our heights were modeled by the Cauchy distribution, there would appear from time to time people who were five, ten, even a thousand meters tall—people who were otherwise just like us except that they just happened to grow that tall, just as Phoebe just happened to stop almost parallel to the wall. In part III of this book, we will see how Cauchy's concept led to the discovery of some strange Wildovian laws.

Nature Abhors a Vacuum

I left unanswered the question whether a genius should be considered a miracle or just someone who is unusually talented. Is he (or she) made of the same stuff as every other gifted individual? To continue our metaphor, is he simply a manifestation of a Phoebe ending up almost parallel to the wall? When my friend Alex said that what is a miracle to me might not seem so to someone much more talented, he might have added, "You, being relatively close to the center, cannot see very far, but someone who landed farther away might, and that person could perhaps see that the genius was created by the same markswoman that created us."

So far, Alex might as well stick to his conviction. He should be pleased that the discovery of Wildovia supports his position: there are phenomena that seem like miracles to ordinary minds but can be explained by Wildovian laws. The idea of Wildovia is acceptable to Alex, since it is a purely scientific concept. He had previously been familiar only with Mildovia and believed in its laws unconditionally, though apparently not enough to be discouraged from trying to get a Hungarian enterprise listed on the Nasdaq.

The discovery of Wildovia has helped us understand that in the parts of the world governed by Wildovian laws, it is to be expected that there will sometimes occur extreme deviations that defy common sense. Previously, with only Mildovian laws to provide an explanation, such occurrences were considered miracles. But in Wildovia, Alex's point of view rules: such deviations are not miracles, and we are guaranteed even larger deviations in the future.

It may not be only the worlds of finance and technology that are better described by the laws of Wildovia, but the world of human ability as well. If that is so, then Alex again may be right, and genius may be nothing more than an extreme manifestation of talent in

Wildovia. On the other hand, science did more than simply delve ever more deeply into the laws of the wild world. It also determined that there will always be phenomena whose explanation is beyond our current understanding.

The idea that "nature abhors a vacuum" is an ancient principle of natural philosophy going back at least to Aristotle, who argued that if an attempt were made to remove all the material from a region of space, the surrounding material would rush in to fill the vacuum. And indeed, in our everyday experience, nature fills in the blanks sooner or later, at least here on Earth, where gravity tends to pull everything downward. It is quite another matter in outer space, where vacuums are the rule rather than the exception. Nevertheless, there are indications that some form of abhorrence of vacuums has validity in quite a wide domain. By analogy, whenever there is a gap in knowledge—something that science cannot explain—that vacuum will be filled, generally with an explanation that calls the phenomenon in question a miracle.

But perhaps some things will always be considered miracles, no matter how far science advances. In this category I propose the genius, who, according to many indications, is in some ways completely different from the rest of us. In terms of our earlier metaphor, it may be that genius is analogous not to our Phoebe's ending up almost parallel to the wall but to something that has nothing to do with shooting. A genius, then, would not be like a distant bullet hole but something entirely different, even if in every other way a genius looked exactly like the rest of the bullet holes. But if a genius is just another flesh-and-blood person, how can genius be something entirely different from talent? I will answer this question and clarify the vague notions introduced here after I discuss "hyperreal" numbers in the next chapter.

tell whether a certain inconceivably strange event is a true miracle or was simply caused by our friend Phoebe, who happened to spin herself very close to parallel to the wall. To clear up the fundamental difference between these two cases, we shall have to investigate the nature of Wildovia in greater depth, and I will return to this subject in part III.

True miracles are something quite different from a markswoman who happens to spin herself close to parallel with the wall and thereby hits an inconceivably distant point. A true miracle is something else entirely, something unique and unrepeatable. Such things occur in Mildovia just as they do in Wildovia, because they are not a result of the usual laws of nature but of something radically different, something that does not contradict those laws but also is not a consequence of them. We will consider these *true miracles* as well. At the end of the next chapter, I will identify four types of miracles. But first we need one more fundamental concept.

3

The Source of Miracles: Gödel's Idea

If you're the smartest person in the room, you're in the wrong place.

This book marks the fourth time I have written about Kurt Gödel's ideas. Each time was for a different reason. Although the mathematical content of Gödel's theorem has not changed, its implications are so profound that I find myself returning to it again and again.

In my book *Ways of Thinking,* I used Gödel's theorem to show that there are inherent limits to what can be accomplished with purely rational thought and that to transcend those limits, some special trick is necessary. That trick turned out to be human intuition. In *Moral Calculations,* I discussed the concept of cooperation. This apparently simple notion proved surprisingly difficult to define— actually impossible. No matter how you define cooperative behavior, you can then use the technique developed by Gödel to construct a multiplayer game in which, if everyone adheres faithfully to the definition of cooperation, everyone will lose, whereas a different, non-cooperating strategy would allow everyone to win. Thus there cannot be an unequivocal, general, and final resolution to the problem of cooperation. In *The Evolution of Money,* I argued that not only was the technique developed by Gödel adequate to deduce his radically new theorem in logic, but it could also be used to describe the core mechanism of Darwinian evolution, which can be applied to biological evolution, to the evolution of ideas (memes), and to economic evolution (money and capital). Thus what Gödel discovered is much more than an elegant and effective mathematical technique; it turns out to be a natural mechanism that underlies a variety of phenomena.

And now Gödel's idea will show us why, in addition to the far-out occurrences of Cauchy's wild world, other kinds of miracles can exist. The Gödelian point of view will also help us, in part IV, to understand why we discuss certain behaviors and attitudes rather than others, and it will help prepare us for the miracles we shall encounter.

Gödel's Theorem

In 1931, Kurt Gödel (1906–1978) published an article in the journal *Monatshefte für Mathematik*. An English translation of the title is "On Formally Undecidable Propositions of *Principia Mathematica* and Related Systems I." Theorem VI of that paper, later to achieve fame as Gödel's first incompleteness theorem, reads as follows:

> For every ω-consistent recursive class κ of FORMULAS, there is a *primitive recursive* CLASS-SIGN r such that neither $\forall(v,r)$ nor $\neg\forall(v,r)$ belongs to Conseq(κ) (where v is the FREE VARIABLE of r).[1]

Gödel originally wrote this in German, but I can assure you it is no more intelligible to the lay reader in that language than in

English. In ordinary language, the theorem says something like the following:

> Every mathematical system that (1) is based on a finite number of axioms (statements accepted without proof), (2) is constructed in a purely formal way, (3) contains an axiom that implies an infinite sequence of natural numbers (zero is a natural number, and every natural number has a successor), and (4) is free of contradiction (i.e., is consistent in the sense that it is impossible to prove within the system both a statement and its negation) is guaranteed to contain propositions that can be formulated precisely within the system but can be neither proved nor disproved.

Gödel's incompleteness theorem came as an enormous shock to mathematicians and logicians. For twenty-five hundred years—ever since mathematics in the modern sense came into being—mathematicians had firmly believed that every mathematical assertion that could be clearly and precisely stated could, sooner or later, be either proved or disproved using the formal methods of mathematical deduction. One had only to be clever enough to find a proof. But Gödel shattered that dream. He showed that there are mathematical assertions that no one, no matter how clever, can ever prove or disprove.

The four conditions I describe above are not at all stringent or abstruse. They apply to most of the mathematics that we use every day. Gödel's theorem thus guarantees that all but the simplest mathematical systems will encounter problems that, though purely mathematical, are unsolvable with that system's methods. Hence the name "incompleteness theorem."[2]

Gödelian Divertimenti

It is tempting to delve more deeply into the surprising consequences of Gödel's theorem. I will try to resist that temptation so as not to lead us too far from the subject of this book. Fortunately, we have Douglas R. Hofstadter's *Gödel, Escher, Bach,* a book that in itself is a minor miracle; not even the publisher expected that this long, difficult, mathematically oriented book would sell millions of copies. We may say, then, that *Gödel, Escher, Bach,* which presents numerous far-reaching consequences of Gödel's theorem and Gödelian thought, was a real black swan. One reason for its success is that Hofstadter manages to deduce Gödel's theorem in simple language, without complicated mathematical formalisms yet with perfect mathematical precision. Doing so, however, takes him about as many pages as this entire book.

I shall not even try to compete with that magnificent achievement. Instead, let me present a few Gödelian divertimenti to illustrate the scope of the consequences of Gödel's theorem and why it has changed our thinking so profoundly. My hope is that these sketches will suggest why Gödel's seemingly esoteric discovery could form the common core of the topics of my three previous books, which were not at all esoteric, and then be pressed into service again to shed light on the source of miracles.

There is a law in a certain village that the man who is appointed to the post of village barber must shave every man who does not shave himself, and he is forbidden to shave anyone who does shave himself. So who shaves the barber? Observe, first of all, that the barber cannot shave himself, because he is forbidden to shave anyone who shaves himself. But if he does not shave himself, he must shave himself after all, because he has to shave every man who does not shave himself. So our problem has a Gödelian flavor: it is a logical paradox with no solution.

Of course, we may easily conclude that the villagers have passed a stupid law, artificially contrived to produce a paradox, but it illustrates, by analogy, the very essence of Gödel's theorem: in every formal legal system, contradictions will exist. From this point of view, it doesn't matter whether a particular law is stupid or clever. Whatever it states, a situation can be constructed in which the law ought to apply but doesn't. And if such a situation can be imagined, sooner or later it will be in someone's interest to create it in order to avoid the law's consequences. So it is that we regularly encounter loopholes, and they are unavoidable: if you add more regulations to close the loopholes, you will only create new ones.

Some time ago, when Hungary was under Communist rule, a number of books were banned by the authorities. It was against the law to sell or distribute such books or even to own or read them. Occasionally I would be consumed by curiosity and try to find out the titles of the banned books. I never succeeded, because the list of banned books was itself banned. The list became a Gödelian concept. This effort at concealment was hardly necessary; the Catholic Church's list of prohibited books—the *Index Librorum Prohibitorum*—which was in existence from 1559 to 1966, was not itself prohibited.

Such is the nature of dictatorship. Extreme totalitarian ideas are not only morally dubious, but also often self-contradictory. These ideas are totalitarian because they promise a complete, formal answer—as if one were dealing with problems in pure mathematics—to important questions of human existence, such as how people can be made happy. But since there are infinitely many value systems and forms of happiness, the spirit of Gödel's theorem applies here as well. No mathematically precise ideology can make everyone happy. In every society, there will inevitably be perfectly normal people who are held back from self-actualization by the social norms of their milieu. If a society claims it has no such individuals, then either it is lying or, by Gödel's theorem, the social system in question is inconsistent. To phrase it slightly pompously, Gödel's theorem guarantees that there is no solution to history.

Those of us who lived through Soviet control of Eastern Europe consider it a miracle that the dictatorships of the twentieth century were eventually overthrown. While they were in power, we could not envision any realistic scenario that would lead to their collapse. We knew, of course, from firsthand experience that they were self-contradictory and therefore ultimately unsustainable, but Gödel's theorem did not predict their collapse. It reveals to us the mechanism by which miracles can emerge, but it does not guarantee any particular miracle.

The theorem can also be applied to the branch of philosophy called aesthetics. Since there is an infinite variety of beauty, Gödel's theorem guarantees that for every consistent system of aesthetics, there exists a type of beauty (and also a type of ugliness) that cannot be deduced from within the system as beautiful. It is no surprise that we encounter so many Gödelian beauties in works of art. Hofstadter presents this mostly through Escher's engravings and Bach's fugues, but one may also find numerous examples in literature.

In one of the short stories in Stanisław Lem's *Cyberiad,* the ingenious engineer Trurl builds a Perfect Adviser for the evil king Mandrillion, who immediately commands the Adviser to get rid of Trurl so that the king will not have to pay the engineer his fee.[3] Trurl wants his fee, but how is he going to get it? If he tries to get the king to pay up, he will have to oppose the perfect mind he created. The Adviser sees through all of Trurl's stratagems and protects the king from every effort by Trurl at extracting payment. Yet Trurl ultimately succeeds. He begins to send the Adviser friendly, innocent-sounding letters. The Perfect Adviser is no dummy, of course, and he figures out that the engineer's gambit is precisely that the letters are meant to arouse suspicion through their seeming innocence. There must be some secret code hidden in those innocent words. Despite the Adviser's protestations to the contrary, the king becomes convinced that there must be some conspiracy between Trurl and the Adviser, and when one of the letters mentions the Adviser's purple screws, of which the Adviser denies all knowledge, the king orders that the Adviser be dismantled down to the last screw. But without his Adviser, the king is vulnerable to Trurl's superior intellect and finally has to pay him.

Trurl sums it up thus: "It was once said that to move a planet, one need but find the point of leverage; therefore I, seeking to overturn a mind that was perfect, had to find the point of leverage, and this was stupidity."[4] Trurl was certain from the outset that Gödel's theorem guaranteed the existence of a point of leverage, but it took the genius of the constructor to find an actual Gödelian question that finally beat the combined minds of the Perfect Adviser and King Mandrillion.

In Jorge Luis Borges's short story "The Lottery in Babylon," the Babylonian lottery is an instrument of fate, and fate has a way of dealing both good and bad.[5] A slave who was too poor to purchase a lottery ticket stole one, and when the lots were drawn, he found that he had won the privilege of having his tongue burned out. But he was also to be punished for having stolen the ticket, and the Babylonian code stipulated that the penalty for such a theft was also to have one's tongue burned out. Thus a conundrum arose: should the slave lose his tongue for the crime of theft, or should he, as his more magnanimous fellow citizens argued, lose it simply because fate had so decreed? There was no simple resolution to this Gödelian problem. If the law in Babylonia had required that one's tongue could be burned out only if the reason for so doing could be determined unequivocally, then for the slave, a miracle would have occurred: he would have been allowed to keep his tongue even though technically it should have been burned out twice.

Gödel's Trick

There is an entire family of jokes that turn on a group of passengers in a train compartment—or sometimes they are patients in a psychiatric clinic or inmates in a prison—referring to jokes by their numbers. In one version, a newcomer to the group randomly calls out a number, and the passengers attack him for having told a terrible joke. In another, they all burst out laughing because it was a joke they hadn't heard before.

Gödel's clever idea was to assign a number to every mathematical statement. It is not exactly something to make you roll on the floor with laughter, but it can be done nevertheless, and once we can refer to statements by their numbers, we are permitted an important degree of mathematical formalization. To number the statements means to list them in some order. We first observe that every mathematical statement can be expressed as a formula—for example, in the system of *Principia Mathematica* mentioned in the title of Gödel's paper.[6] So we begin with the statements consisting of a single symbol, and when we run out of those (which we will, because such a system must have only a finite number of symbols), we then proceed to all statements consisting of two symbols, and so on. It should be clear that sooner or later, every possible statement will make the list and hence get a number attached to it. The Pythagorean theorem will have a number, as will the statement "2 + 2 = 4" and the theorem on the factorization of the difference of two squares: $a^2 - b^2 = (a + b)(a - b)$. And, of course, every false statement will also have a number, such as the statements "2 > 3" and "2 + 2 = 5" and the incorrect expansion $(a + b)(a + b) = a^2 + b^2$.

Gödel then went a step further and numbered separately all valid proofs. Just as was done for statements, a proof that establishes the validity of a mathematical statement can be represented as a series of logical formulas that obey certain rules. Gödel used the same method that he had used with formulas: he began with the one-symbol proofs, proceeded to two-symbol proofs, and so on. The result was that every possible valid deduction now had a number denoting its position in the sequence of all well-formed proofs. Since the proofs are ordered by length, every proof, no matter how long, will eventually appear in the list.

This description somewhat simplifies what Gödel actually did. For technical reasons, he used a much more complicated system to number the formulas and proofs. But what I have presented here captures the basic idea. All of this numbering of statements and proofs had the sole purpose of guaranteeing the existence of a very strange statement in his list—a statement that was later given the name *G* in Gödel's honor. Translated from mathematical symbols into English, statement *G* reads as follows: *There exists no natural number x such that the proof number x is a proof of G.* In other words: *The list of all possible proofs does not contain a proof of the statement you are reading right now.*

Gödel's master stroke was to phrase this strange, self-referential formula in a mathematically precise way. He then went on to prove that the statement *G* cannot be proved (that is, that there is no proof of *G* in his list). And its negation cannot be proved either (because, as we will see, it is in fact false). If *G* were one of the numbered jokes told by our railway passengers, the passengers could argue until the end of time whether joke *G* should be laughed at or panned, because there would be no possible proof one way or the other.

Mind you, *G* is obviously a "good" joke in the sense that it is true, for if it weren't true, then its negation would have to be true. There would exist a natural number *x* such that proof number *x* proves *G*, but since *G* says there is no such number, that would mean that *x* proves its own nonexistence. That means *G* must be true—but if that's so, then it must be possible to render this paragraph as a finite set of mathematical symbols, which would give it a place on the list, which would mean that *G* must be false. Who shaves the barber?

Mathematicians, unlike our jokesters, did not become confused and argue endlessly. They learned to accept that this is just how mathematics works. In fact, many problems that had remained unsolved for years have turned out to be statements that cannot be proved or disproved, so it is no surprise that no one could solve them.[7]

Hyperreal Numbers

Decades later, it occurred to the American mathematician Abraham Robinson that it would be interesting to add the negation of *G* as a new axiom to the classical system of mathematics.[8] After all, he reasoned, it would still be a mathematical system, and if classical mathematics is consistent—that is, there is no statement such that both it and its negation can be proved—then the mathematics obtained by adding this one axiom would also be consistent. If the new system were inconsistent, one would be able to prove both *G* and not *G*, but since the only difference between the old and new systems is the addition of the axiom not *G*, which cannot be used to prove *G*, it follows that *G* can be proved in the new system only if it can be proved in the old system, which, as Gödel proved, it cannot. If the new system were inconsistent, one would be able to prove both *G* and not *G*, but since *G* cannot be used to prove not *G*, it follows that not *G* could have been proved in the original system. Hence, if we append the negation of *G* to classical mathematics, we get a consistent mathematical system, on the assumption, of course, that classical mathematics is consistent to begin with.

But it is far from obvious that classical mathematics is consistent. Although most mathematicians are convinced that it is, no one has succeeded in proving this. They know, however, that it cannot be just a tiny bit inconsistent. It was proved hundreds of years ago that if classical mathematics were to contain a single contradiction, then for every statement that can be proved, its negation could be proved as well. Therefore, mathematics is either perfectly consistent or else riddled with contradiction. This is as good a reason as any for mathematicians to have faith in its consistency. Gödel caused a shock in this respect as well, because it follows from his theorem that it is impossible to prove the consistency of any sufficiently rich system of mathematics within the system. Consistency will always remain in some sense a question of faith.

In light of this fact, Abraham Robinson's idea seems absurd. After all, he decided to construct a mathematical system containing an axiom he knew to be false. It is as if I were somehow not only the man I am but a woman as well. Of course, in real life, this is impossible (ignoring, for the sake of this example, hermaphroditism and bigenderism). I am not a woman, and there is no one out there that is both I and a woman. But such paradoxes are possible in mathematics. If classical mathematics is consistent, then the new system will also be consistent, since Robinson is adding an independent axiom. Mathematically, this new system will be as clean and orderly as the old one. So there is no harm in examining the new system to see what new theorems can be deduced within it.

It turns out that if we work in this new system, the meaning of being the number of a proof in the sense of Gödel's numeration will have to change. Adding the negation of G required the addition of a certain "generalized" natural number. Hofstadter calls it a "supernatural number," since there is something almost miraculous about it. To give our supernatural number a name, let us call it I, since it is a number from our imagination. All of traditional mathematics will continue to get along just fine without I, while all mathematics involving the negation of G will have to use I. If a calculation involving I produces a result in traditional mathematics, the I will disappear, just as that other mathematical i, namely the square root of -1, disappears in certain calculations, for example, $(1 + i) \times (1 - i) = 2$.

Harry Potter aficionados will recall that Platform 9¾, which leads to Hogwarts, is also an imaginary platform of a track that nevertheless leads somewhere.

Robinson generalized the notion of supernatural numbers to the entire set of real numbers and called them "hyperreal numbers." This supernatural number I cannot be any of the traditional natural numbers, since that would make G provable in classical mathematics. So I cannot be equal to 0, 1, 2, 3, or any other natural number. What works is to consider I to be greater than every natural number yet still an actual number, not some sort of infinite quantity. It has a double and a square, which are also hyperreal numbers. Two hyperreal numbers can be added and subtracted, and hyperreal numbers can be combined in all the usual ways among themselves and with the natural numbers as well. In this new mathematical system, for example, $2I$ and $I + 3$ are valid numbers.

Recall that the hyperreal numbers arose from adding the negation of G as an axiom to traditional mathematics, and since that axiom was independent of traditional mathematics, we can safely perform all the usual calculations with both real and hyperreal numbers without worrying about arriving at a contradiction (assuming, as usual, that traditional mathematics is consistent).

Now what about the number $1/I$? Since I is greater than all the natural numbers, it follows that $1/I$ must be smaller than all the positive real numbers, while itself remaining positive. That is, our system contains positive numbers that are "infinitesimally" small. This happens to be precisely the sort of number mathematicians have been talking about since the development of calculus by Newton and Leibniz. For those men, it was more an abstraction than a concrete object. But now it is here, right in our hands, a "genuine" infinitesimally small number! And all on account of Gödel's theorem. What Robinson did with these infinitesimals was to create a new version of the differential and integral calculus called nonstandard analysis.[9] It turned integration and differentiation from esoteric concepts that have tormented generations of first-year calculus students into trivial operations—the complex operations involving limits become simple calculations with infinitesimal numbers.

But there's a catch: in this system, addition and multiplication become extraordinarily complicated. It has been proved that if you want to do nonstandard mathematics with hyperreal numbers, then either addition or multiplication will be as complicated as the calculus operations of integration and differentiation in our traditional mathematics. And if you are using the new nonstandard mathematics, you cannot try to cheat and use traditional mathematics for addition and multiplication, because you never know whether the numbers you are working with are real or hyperreal.

Several works of science fiction posit that humans will use mathematics to prove to some alien civilization that we are intelligent beings. But what if the mathematics that developed in that particular civilization is, from our point of view, a nonstandard mathematics? The aliens we are trying to contact might see even integers in a completely different way. Perhaps their youngest offspring readily solve problems in differential calculus, and our mathematics might convince them only that we are hopelessly underdeveloped, with an appallingly meager concept of numbers. They might be shocked and dismayed to see that we have difficulty solving even the simplest equation of motion. We, meanwhile, might feel superior when we discover that the aliens can barely add two ordinary ten-digit numbers.

If this alien civilization, with its alien mathematics, were to create some technical accomplishment that was beyond our power, such as a time machine, we would almost certainly consider it a miracle. If the richness of the mathematics leading to such an invention were inconceivable to us, it would look to us like magic. Then again, perhaps our aliens, hindered by their difficulties in addition and multiplication, never discovered the laws of electricity. They might look on the electric motor as nothing short of a miracle.

Let us bring back our markswoman Phoebe and see what she can do with this new number system. If we confine ourselves to the traditional real numbers (both integers and nonintegers), then we are able to express every possible result of a rifle shot in the real world. But if we expand our horizon to allow for hyperreal numbers, then a shot has other possibilities. For example, a shot might land at a hyperdistant point $I + 3$ kilometers from the center of the wall, although the probability of such a shot might be an infinitesimally small value like $1/I$. But note that this tiny probability is not zero: $1/I$

emphatically does not equal the real number zero. It is greater than zero, even though it is smaller than every positive real number. So in a certain sense, we can think of such a probability as "essentially" zero. Once we allow hyperreal numbers into our midst, we have results that are so impossibly rare that we can never anticipate their occurrence. Yet they *might* occur, and if one does, it will be completely different from an everyday, garden-variety shot. To those of us raised on the traditional real numbers, such a shot, made possible by Gödel's theorem, would seem like nothing short of a miracle.

Gödel, Einstein, and von Neumann

In describing John von Neumann, the historian of science Jacob Bronowski wrote, "He was the cleverest man I ever knew, without exception. He was a genius."[10] Even when he was still a student, von Neumann's high school classmates at the legendary Fasori Lutheran High School, in Budapest, knew he was a genius, and they weren't just anybody. That school produced a number of world-famous scientists, including Eugene Wigner, John Harsanyi, and Edward Teller. Von Neumann died young, at age fifty-three, leaving behind trailblazing work in subjects that included computer architecture, quantum mechanics, and game theory. In the years since his death, five people have received the Nobel Prize in Economics for results in game theory, and another ten or twelve Nobel Prizes have been awarded to economists who applied mathematical techniques that von Neumann developed.

Von Neumann was not the absentminded, head-in-the-clouds genius of popular imagination. There was nothing otherworldly about him. His thoughts could soar while both his feet remained on the ground. When in the 1930s, the newly founded Institute for Advanced Study in Princeton, New Jersey, made him an offer of membership, he demanded a $16,000 salary, a handsome sum back then. And he had a further condition: he did not want to find himself in a place where there was no one smarter than himself. He believed that if you're the smartest person in the room, you're in the wrong place. Fortunately, Albert Einstein and Hermann Weyl had already been recruited to the institute, and Kurt Gödel would join two years later. All three men were refugees from Hitler's Europe.

Von Neumann asked Einstein what sort of salary he was thinking of requesting. Einstein modestly replied that he thought he might be worth a few thousand dollars a year. Von Neumann then ordered Einstein to disappear for a few days, during which time he negotiated an $18,000 annual salary for him.

Einstein and Gödel often took daylong trips to the woods around Princeton. Occasionally, von Neumann or another scientist would join them, but mostly it was just the two of them. On one of those trips, neither of the men spoke a word the whole day long, and when they got home, each told his wife that they had had the most enchanting conversation. It turns out that it matters with whom you share your silence.

Was von Neumann a genius? Most mathematicians would probably say yes. I can best illustrate his extraordinary talent with an actual example. A problem appeared in the mathematical folklore in the 1940s that was quite challenging: how thick would a coin have to be so that when tossed into the air, it would land heads, tails, or on edge with equal probability? Obviously, an everyday coin lands on its edge extremely rarely, but if we start thickening it, the odds improve. If, for example, the coin looked like a can of soup, it would fall on its edge much more often than it landed heads or tails. So there must be a sweet spot between regular coin thickness and soup-can thickness where the probabilities of heads, tails, and edge are precisely equal. Where is it?

I encountered this problem in college as a third-year mathematics student. The obvious approach involves calculus, but the computations are fierce, and it took me a good two weeks before I had finished calculating all the necessary integrals. (The answer is that the ratio of the thickness of the coin to its diameter should be 0.354.) Legend has it that the problem was presented to von Neumann at a party in the United States. After hearing it stated, he stared into the middle distance for half a minute or so, and then announced the answer. The partygoers grew excited, for surely Johnny von Neumann had found an elegant solution to this beautiful problem that might be understandable even to the non-mathematically minded. They turned to him in great expectation. How did you do it, Johnny? Von Neumann replied simply, "Well, I just computed the integrals."

Von Neumann, like Gauss, was not a genius by our standard; I would call him an extraordinary talent. His intelligence is the kind that happens when Phoebe spins herself nearly parallel to the wall and shoots almost unimaginably far, but only almost. His almost unimaginably rapid calculation of integrals shows a mind able to do ordinary things extremely fast, not one that conceives ideas that were previously inconceivable. Von Neumann may have seen himself that way too; he considered Gödel and Einstein more intelligent than he was. Note, however, that the story about Einstein's salary suggests that in commonsense matters, von Neumann was much smarter. On the other hand, Gödel and Einstein, who were geniuses by any standard, had thoughts of a sort that never crossed von Neumann's mind. Another testament to von Neumann's intelligence is that he was one of the first mathematicians to appreciate the magnitude of Gödel's theorem, and he immediately abandoned his researches in mathematical logic.

Both geniuses paid a price. A number of psychologists see in Einstein a possible diagnosis of Asperger syndrome, a mild form of autism. Gödel was clearly paranoid. For example, he was convinced that his food was being poisoned, and he would not eat unless his wife or Einstein tasted his food for him. When his wife was hospitalized—long after Einstein's death—he starved to death at age seventy-two.

I have no intention of reinforcing the stereotype of the genius as nut case. There is no scientific evidence that all geniuses are mad. To establish such a claim, we would first of all need a much larger sample of geniuses. Perhaps Newton is a counterexample, though he too is often given a posthumous diagnosis of autism, or at least Asperger's, or sometimes bipolar disorder or paranoia.[11] With such diagnoses falling all over the map, it might just as well be that Newton was altogether healthy, only inconceivably intelligent, of course, so that he must have stuck out from the throng. He held important positions of trust, including the Lucasian professorship of mathematics at the University of Cambridge and the presidency of the Royal Society, and he was warden and master of the Royal Mint.

The tales of Einstein and Gödel and how they differed from John von Neumann do not themselves prove that the mind of a genius is qualitatively different from minds of extraordinary talent. Nonetheless, these stories show why I think genius is a genuine miracle, the protests of my friend Alex notwithstanding.

But perhaps Alex is right after all. If we allow for the existence of hyperreal numbers, then a genius is an offshoot of the same marksman who produced more ordinary minds. But in the case of a genius, our Phoebe made a shot far beyond far, a shot whose probability was so infinitesimally small that it seems like zero to those of us whose minds were nourished on traditional numbers.

The Four Types of Miracles

It would be foolish for me to continue talking about miracles without trying to define what I mean by such a thing. We human beings are creative users of language, and we use words not only in their literal meanings, but also figuratively. After a while, a word takes on a variety of meanings. Look up "miracle" in Webster's and you will find the following:

0. A wonderful thing worthy of admiration: a truly superb representative of its kind.
1. An accomplishment or occurrence so outstanding or unusual as to seem beyond human capability or endeavor.
2. An event or effect in the physical world deviating from the laws of nature.
3. An extraordinary event manifesting divine power or intervention in human affairs.

These four quite different definitions appear here in order of increasing miraculousness. Let us put aside definition 0 at once. I am not interested in such "miracles" as the prize-winning cheese at the county fair or even Mr. Micawber's snatching of a crucial letter in Charles Dickens's *David Copperfield:* "Uriah . . . made a dart at the letter, as if to tear it in pieces. Mr. Micawber, with a perfect miracle of dexterity or luck, caught his advancing knuckles with the ruler, and disabled his right hand."

Let us call occurrences that satisfy definition 1 *pseudomiracles.* They represent an enormous deviation from the mean, of the sort that arises from time to time in Wildovia. These are not miracles by our definition because they recur in a statistically predictable way. They are simply rare events. And under the laws of Wildovia, such as those arising from the properties of the Cauchy distribution, any such miracle will eventually be superseded.

In defining *true miracles,* I want to modify definition 2 to state that miracles encompass phenomena that deviate from the laws of nature *as currently understood by science.* Gödel's theorem—at least in spirit, if not in its formal mathematics—convinces us that such occurrences will always arise, regardless of how far and in what directions science advances.

Finally, let us call phenomena that satisfy definition 3—those caused by divine intervention—*transcendent miracles.* They will never be amenable to scientific explanation, for the existence of such miracles will always be a question of faith.

Pseudomiracles may not be "real" miracles, but they can certainly seem miraculous. To me, it seemed a miracle that John von Neumann needed only thirty seconds to calculate in his head what had taken me, an International Mathematics Olympiad medalist, two very laborious weeks with pencil and paper. Even if it really wasn't half a minute—no one was timing him with a stopwatch; it may well have been a minute or two—it was still an astonishing feat. Such is the chasm between a merely gifted mathematician and a talent so enormous that only one or two of their like are born in a century. We have good reason, therefore, to consider a mathematician of the class of von Neumann, Gauss, or Cauchy a miracle in the everyday sense. But such miracles are explained fairly well by the science of Wildovia.

True miracles are different. They defy explanation by the science of their time, just as today's science cannot precisely explain the collapse of the Soviet Union or the fall of the Berlin Wall. But today, with Gödel's theorem under our belts, we can say that such miracles can occur at any moment, in both Mildovia and Wildovia, and take us quite by surprise. Therefore, it is worthwhile to prepare for them and to gather experience on how we might—even if we cannot predict them—at least adapt to their existence, prevent them from causing excessive harm, and exploit them if they happen to be positive.

My cousin once came down with a nasty skin infection. Her father, a professor of medicine, consulted a colleague, a famous dermatologist, who took her under his care and treated her with the greatest possible attention, as became the child of a colleague, exploiting all the latest accomplishments in medicine. Months passed, but the infection would not go away. One evening, during a family dinner, my father posed an unexpected question to his physician brother: "How would you treat your little girl if you were a family physician in some tiny village way off on the Slovenian border?" My uncle replied at once, "I should, of course, apply a chamomile poultice." So he tried that, and my cousin was fine in three days. According to contemporary medicine, we had witnessed a true miracle, even if a hundred years ago it would not have been any sort of miracle at all, because back then they would have applied a chamomile poultice in the first place and not been surprised that it worked. And perhaps it will not be a miracle a century from now, because by then, medicine may have understood that type of skin infection and its response to chamomile.

Transcendent miracles are honest-to-God miracles in the strongest sense of the word. There is no science, mathematics, or statistics to explain their existence, although for a person of faith, such divine manifestations seem explainable enough. I, however, being of a scientific frame of mind, shall not distinguish such manifestations from what I am calling true miracles, and I shall leave aside entirely the question whether there are in fact transcendent miracles or whether science will eventually be able to explain every single miracle. One thing is certain: science will have a hard time explaining genius.

These four categories distinguish only among the basic definitions of miracles; they say nothing about the defining characteristic of miracles—that is, that they are unique and unrepeatable. We will return to this topic in part III, after we have seen that several phenomena of Wildovia, while not guaranteed to be unrepeatable, are at least guaranteed to be unpredictable. The most intriguing thing— the miracle, if you will—is that these results will emerge not from some specialized science of miracles but from the normal, completely miracle-free, Mildovian scientific method.

PART TWO
The Mild World

There is more of value to be found in a rich man's junk heap than in all of a poor man's possessions.

4

The Power of the Normal Distribution

Some laws of nature steer the world toward Mildovia; others, toward Wildovia.

Sir Francis Galton (1822–1911), Charles Darwin's cousin, was a true polymath. He discovered the meteorological phenomenon known as an anticyclone; he was the first to suggest the use of fingerprints for identification; he was the first to chart certain parts of Namibia. He examined whether prayer lengthened human life and could demonstrate no such effect. He was also the first to apply the mathematical concepts of the average difference from the mean and standard deviation to psychology.[1] He needed those notions because although he studied the world of Mildovia, he wasn't interested in the average so much as the extremes. For example, he wanted to discover the extent to which extraordinary talent might be hereditary.

Galton examined the children of parents who were far above average in certain measures to see how they turned out. At the time, standardized intelligence tests and other psychological measuring instruments hadn't been invented, so he used a qualitative idea of excellence. His category of "far above average" included judges, scientists, leading politicians, and famous physicians, among others. In the Victorian era in which Galton lived, such professions were open only to men, so Galton studied fathers and sons. He found that the son of an outstanding father was, on average, above average but generally not as outstanding as his father. Why should that be? Does it mean that the world tends toward mediocrity? Perhaps it is so in Mildovia, where phenomena tend to cluster around the mean. But even in Mildovia, it turns out that there is no general trend toward the average.

Of course, sons have mothers as well as fathers, and perhaps it was they who were responsible for the diminished quality of the sons. So Galton extended his research to domains in which there is only a single parent. Instead of studying fathers and sons, he began analyzing the offspring of tobacco plants propagated asexually. He selected a characteristic that was much easier to quantify than human excellence: the length of the leaves. Galton found that the leaves of the offspring of long-leaved tobacco plants were also longer than average, but in general not as long as the leaves of their parents.

Regression toward the Mean

It would seem, then, that it is not sexual reproduction that causes the sons of outstanding fathers to be less outstanding, but that doesn't make the observation any less mysterious. Galton called this phenomenon "regression toward the mean"—in other words, a return to the average. But a label is not an explanation. Why did the entire population not become more average over time? For in fact, it did not. Galton examined several generations of tobacco plants, and he found that in each generation, not only was the average leaf length about the same, but the standard deviation was the same as well: each generation contained about the same proportion of extraordinarily long-leaved plants. So the population of tobacco plants appears not to be slouching toward mediocrity, and neither are humans; there are outstanding individuals in every era.

Galton reasoned that if the children of extraordinarily talented individuals did not provide the next generation of extraordinary talent, it must be that the children of average, or perhaps slightly above average, parents would somehow turn out to be outstanding. So he decided to run his previous study in the other direction; this time, he would examine not the children but the parents of extraordinary talents. Of course, there could be no question here of causation; while a father might or might not influence his son's accomplishments, it was absurd to think that a son's talent could somehow be transferred retroactively to his father. Galton's cousin Charles Darwin had in any case done away with the concept of improvement in biology. The evolution of organisms is governed by chance suitability to an environment, not by development toward some "improved" form. In light of this discovery, the idea of looking in the other direction became less absurd.

Since human intelligence is so complex, Galton again decided to study a variable he could measure precisely. He performed his reverse experiment on tobacco leaves. His results were fully commensurate with those of the first experiment: the ancestors of extraordinarily long-leaved plants usually had shorter leaves than their descendants, though still longer than average. He was therefore not surprised when, turning his attention again to human beings, he found that the fathers of outstanding men were usually above average, but by and large not nearly as much so as their sons.

The phenomenon of regression toward the mean also appears at the other end of the scale. Extraordinarily low intelligence slowly returns to average over generations, and at the same time, extremely low intelligence is seen in the offspring of average or only slightly below-average parents.

Regression toward the mean is, then, a purely mathematical phenomenon that will always appear if we examine two separate variables by determining both of their values for every individual in a population. It doesn't matter what the variables and the population actually are. This type of regression is a quality of the mathematical structure used to analyze a population, not of the specific properties of human beings, tobacco leaves, or inheritance.

I can illustrate this with an extreme example. Suppose I claim I am a powerful wizard who can reverse misfortune with a dab of incantation. Behold as I demonstrate my ability! I ask a thousand people each to throw three dice. On average, about five people will make an unlucky roll—let us say three ones. These are clearly unlucky individuals. But I can banish their bad luck: I chant my magic spell and let them throw again. It is highly improbable that any of the five will again throw three ones. I pronounce them cured! In fact, however, they are simply subject to the mathematical law of regression toward the mean.

This example is extreme because there is no connection whatsoever between the first and the second sets of dice rolls. Clearly, the second roll does not "inherit" the results of the first roll. If it did and the inheritance were perfect, then whoever rolled three ones the first time around would throw them again, no matter what magic words I might mumble.

Biological inheritance generally falls within these two extremes: characteristics are passed on to offspring to some degree but with nowhere near one hundred percent certainty, even in asexual reproduction. Regression to the mean is always there but never as starkly as in the example of the unlucky dice throwers. The stronger and more precise the inheritance, the smaller the effect of offspring regressing to the mean, while a weaker effect of inheritance leads to stronger regression.

The mathematical phenomenon of regression toward the mean does not even have to be related to inheritance. "Regression" is in fact a set of statistical techniques for estimating the value of one variable from those of another variable. It makes a difference whether we are estimating the value of variable *A* from values of variable *B* or values of *B* from those of *A*. If, for example, we ask for the average height of men weighing two hundred pounds, the answer is around six feet. But if we ask for the average weight of six-foot-tall men, the answer is not 200 pounds or even 190 pounds. In this case, we may not be that surprised, since a very short person can weigh hundreds of pounds, bringing down the average height. But sometimes the results are rather counterintuitive. For example, if we try to predict when the human population of the Earth will reach ten billion, the two variables in question are population and time. Let us imagine that according to all available data, the most likely estimate is the year 2050 for a population of ten billion (estimating a time value from population values). But if we ask how large the Earth's population will be in the year 2050 (estimating population from time values), the most accurate answer according to the same set of data is 9.3 billion. And these different answers are not even contradictory.

Reward and Punishment

Psychologists often investigate the effects of reward and punishment. Several studies have shown that performance tends to weaken after a reward is given and becomes stronger after a punishment. Thus many have drawn the conclusion that punishment has a positive effect and reward a negative one. But this conclusion is false, for it ignores regression toward the mean, which does not arise out of the particular nature of the reward and punishment. It comes from the way we examine and analyze our data. With this in mind, the correct conclusion might be the exact opposite.

People are usually rewarded when they perform better than usual and punished when their performance becomes worse. Regression toward the mean tells us that without any reward or punishment, an unusually outstanding performance will generally be followed by a weaker one, and an unusually bad effort will likely be followed by a better one. If we don't take regression toward the mean into account, we cannot draw any conclusions about the effects of reward and punishment.

We must compare the degrees of improvement and deterioration without any extrinsic effect (reward, punishment) with what we see when we do give rewards and punishments. If the deterioration following an outstandingly good performance is on average greater without a reward than with one, then reward must have a beneficial effect. Similarly, even if performance generally improves after a punishment was administered, the punishment might still be depressing performance, since without it, the average improvement might have been greater.

Had we not known about regression toward the mean, we might have been fooled into thinking that rewards are bad and punishments good, since every piece of research data pointed in that direction. In fact, the picture is a good deal more complicated. In some situations a reward can be beneficial, while in others it can be harmful. The same holds for punishments. Not only the situation but also the individual's personal qualities determine the overall effect of a reward or a punishment. There are even people on whom they have little or no effect—who can truly say the following of themselves, in the words of William Ernest Henley from his poem "Invictus":

> It matters not how strait the gate,
> How charged with punishments the scroll,
> I am the master of my fate:
> I am the captain of my soul.[2]

Diversity and Stability

Why, despite these complications, does regression toward the mean not lead to the whole population's becoming uniform or at least more and more average? It might appear that it should, but the facts show otherwise.

The lengths of Galton's tobacco leaves followed an unvarying distribution for generations. In fact, this distribution followed with great accuracy the one Gauss had described only a few decades before Galton's investigations. Galton found that if a population is normally distributed with respect to some trait, then it follows, strictly mathematically, that the phenomenon of regression toward the mean will be counterbalanced by the fact that outstanding individuals will be found among the offspring of individuals who are about average. Both the variety and the stability of the population are preserved even in the presence of regression toward the mean, because that is what is guaranteed by the mathematical characteristics of the Gaussian distribution.

Later, mathematicians discovered that the relationship Galton found between variety and stability occurs only if the distribution of the population is approximately Gaussian.[3] Hence, we can look at the normal distribution as a source of stability: it is what enables a population to remain much the same generation after generation, despite regression toward the mean.

Of course, populations do change over time. For example, the human race has become substantially taller in the past century. That is due in part to advances in medicine, in part to the fact that malnutrition has diminished greatly. Nevertheless, it is already clear that this growth has come to a stop, at least in the developed world, over the past few decades. The height of future generations will probably be much the same as the height of our generation, in both its mean and its standard deviation. A new stable height regime has been established. The Gaussian curve has shifted, but it is still a Gaussian curve.

The law of regression toward the mean holds in both Mildovia and Wildovia, but since the stability of a population can be guaranteed only by curves like the Gaussian, stability is restricted to Mildovia. That is why we shouldn't look down on traditional Mildovian science, even if it cannot adequately describe (or model) certain phenomena. Deep down, we all long for stability, and some populations have achieved it. Cockroaches and rats have remained stable for millions of years, maintaining such characteristics as the ratio of larger individuals to smaller ones and lighter individuals to darker ones. Some real-world phenomena are modeled rather well by the laws of Mildovia.

Stability in Mildovia does not imply that a kind of stability cannot exist in Wildovia. Heraclitus of Ephesus said (or is said to have said) that nothing is permanent except change. This twenty-five-hundred-year-old adage aptly describes the forms of stability that exist in Wildovia, a place where the stability of a population of cockroaches is inconceivable. But although Wildovia is wild, it has laws of nature too. Some laws of nature steer the world toward Mildovia;

others, toward Wildovia. The world we live in is formed from the combined effects of these profoundly different types of natural laws. For the time being, let us focus on Mildovia. We will discuss Wildovia below.

The Bean Machine

Francis Galton invented a device now known as the Galton box (figure 6), also called the bean machine, to demonstrate a well-known law of probability. Balls (or perhaps beans) are dropped into the hopper at the top, and on the assumption that a ball jumps with equal probability to the left or right as it drops down one level and hits the pin below, the balls will fill the bins according to what is known as

Figure 6. The bean machine (drawing by Vera Mérő).

the binomial distribution. As more and more balls are dropped, the curve they form better and better approximates the Gaussian distribution. In 1920, George Pólya published an article proving mathematically that this is so; he called his theorem the *central limit theorem*. The term "central" reflects its role in probability theory. As we shall soon see, it also sheds light on one of nature's important tricks that happens to steer the world toward Mildovia.

It is easy to understand why many more balls land in the middle bins than in the far left and right ones. To land in the middle, a ball has to jump three times to the left and three times to the right. This can be done in a number of ways—for example, once to the left then twice to the right, then twice to the left and once to the right (LRRLLR). Another possibility is (LLRRRL), and so on. There are twenty such combinations altogether. But to land in the far left-hand or far right-hand bin, the ball has to jump the same way every time, six times to the left or six times to the right, and there is only one way to accomplish each of these drops. We would therefore expect, after a large number of balls have been dropped, to find about twenty times as many balls in the center bin as in each of the left- and right-hand bins.

It is harder to see why the binomial "curve" produced by the balls should approximate the Gaussian distribution. Why not the Cauchy distribution instead or some other distribution I haven't mentioned? The reason lies deep inside the central limit theorem.[4] Unlike Phoebe's bullets, which followed the Cauchy distribution, the balls in the Galton box line up obediently along the Gaussian curve, with no large deviations. If we built a really big board, say with a hundred rows and columns, and dropped a thousand balls every second, we would expect to wait billions of billions of years before a single ball landed in bin 1 or bin 100. Phoebe will much earlier have fired a bullet into the wall at a corresponding distance from the center.

The central limit theorem has a particularly nice interpretation in biology.[5] Suppose a certain biological trait (such as height) is defined by a number of minor components, each of which can have one of several values, and that we can model the trait as the sum of the components' individual contributions. In this case, the central limit theorem guarantees that the distribution of this trait over a large population will follow a Gaussian curve. This is precisely what the bean machine illustrates. Suppose there are sixty genes that influence height and each comes in two varieties, *tall* and *short*. The more *tall* genes an individual has, the taller that individual will be. If you think of a *short*-valued gene as a jump to the left and a *tall*-valued gene as a jump to the right, then to be maximally tall, an individual would have to receive all sixty *tall* genes, a result that is mathematically equivalent to making sixty rightward jumps in a row. Similarly, a maximally short individual would have received all sixty *short* genes, equivalent to sixty leftward jumps in a row. Most individuals will have a mixture of *tall* and *short* genes, in the same proportions as the balls in the bean machine's bins.

To be sure, the components that determine height may be environmental as well as genetic. Several genes provide contributions to our body height, all of them relatively minor. And there are environmental factors as well, such as childhood diet. The case is similar with intelligence if we adopt a simplified model, though even then few genetic factors have been identified, and the environmental factors are also very diverse, ranging from a child's nutrition level to how he or she is read to and spoken to. Nobody is allotted every one of the factors contributing to greater intelligence, but those who are allotted more will presumably be more intelligent. Again, this outcome is precisely what the bean machine models.

Of course, biological phenomena are much more complex than the bean machine. The events that contribute to a ball's final resting place—jumping to the left and jumping to the right—are independent of one another. Whether a ball jumps to the left or right at a given level does not depend on whether it jumped left or right at the previous level. This independence is what leads the machine to approximate the Gaussian distribution.

Such independence rarely holds in biological systems. Each factor contributing to a given trait, such as a gene or an environmental influence, is generally not independent of the other factors. Moreover, the different factors often have different degrees of influence on the final trait. So in fact, the bean machine models the world of living things only to a very limited extent.

Stability through Multiple Components

Mathematicians haven't taken any of this lying down. Many types of central limit theorem have been proven, showing that variants of the binomial distribution also approximate the Gaussian distribution. For example, it has been shown that the components contributing to a trait do not all have to have the same magnitude. On some levels of the bean machine, they can jump two or three columns to the right or left rather than just one. This scenario is more difficult to realize physically, but the mathematical result still holds: the beans in the bins will eventually sort themselves along a Gaussian curve. Events at one level also do not need to be entirely independent of those at the next level. For instance, a ball at one level might be affected to a certain extent by how it jumped on the previous level.

Other variants of the central limit theorem are still being discovered. The overall picture based on several variants of the central limit theorem can be summed up roughly as follows:

- If a trait is defined by several minor components ("minor" in the sense that no single one dominates the rest),
- and if those components are not highly interdependent (a few of them do not determine the values of all the others),
- then that trait will be distributed over the entire population according to the Gaussian distribution.

The distribution of IQ scores in figure 7 illustrates this picture precisely. It looks very much like the Gaussian curve we have been looking at, except for a small bump around the value 70 that mars the otherwise smooth picture. That bump represents the population of individuals with Down syndrome.

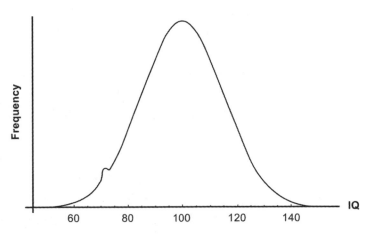

Figure 7. The distribution of IQ scores (drawing by József Bencze based on data in Kun and Szakács, 1997).

Down syndrome is a genetic disorder caused by the presence of an extra twenty-first chromosome. This extra chromosome's effect on intelligence more or less dominates the effects of all other components. It doesn't dominate them to the extent that the individual is completely intellectually disabled, but most adults with Down syndrome have an IQ between 50 and 70, and all other factors determining intellectual development contribute very little to change this figure. In the case of intelligence, any other such dominant component is so rare that its effect would not show up on the distribution curve (the incidence of Down syndrome is about one per one thousand births).

Height follows roughly the same pattern. There are a few genetic components that inevitably lead to unusual shortness or tallness, so that while the distribution of height follows the Gaussian curve quite well, there are a few bumps. Body mass, on the other hand, is different. We shall see why in the next chapter.

The mathematics of the central limit theorem is a foundation on which nature can build stable structures such as the populations of living organisms. Perhaps as nature was experimenting throughout the universe, trying this and that, she came up with both stable and unstable structures. By definition, it is the stable structures that survived. From what we know of the physics that formed the cosmos following the Big Bang and the laws of biological evolution, an effective way to achieve stability is to arrange things so that a number of more or less independent components of comparable importance combine to establish a particular trait. For such a combination will guarantee that the given trait will be approximately normally distributed, a factor that in itself guarantees stability from generation to generation (unless environmental conditions change suddenly).

But try as she might—or perhaps it is just in the nature of things—nature sometimes cannot organize a trait as an interaction of many small components. Sometimes, as in the example of Down syndrome, a component will emerge that dominates all the others. But to the extent that no such dominant components arise, nature's trick of crafting every important trait as the sum of several minor, more or less independent components is usually enough to ensure stability.

I don't know what nature's actual governing principle is, to strive for stability or simply to assemble everything from numerous minor components, with stability a by-product of the construction method. In any case, the central limit theorem tells us why so many things in nature work according to the laws of Mildovia, why Wildovian instabilities do not prevail everywhere. Recall from our discussion of the Cauchy distribution that where Phoebe's shot hit the wall was determined by a single component, namely, her angle to the wall following her spin. At those times when she ended up nearly parallel to the wall, minute differences in the angle would lead to enormous differences in the result. We should not be surprised, then, that the result is Wildovian—unstable in the traditional sense of the word. In Mildovia, where phenomena result from the interaction of many small components, we expect stability, with a well-defined notion of the average, or mean, and a well-defined notion of standard deviation from the mean. But in Wildovia, the only thing that is normal is abnormality. Anything goes, and events don't have a standard deviation.

Here is a fundamental difference between the two worlds. Things that can be described with the aid of the Gaussian curve, which is the very foundation of Mildovia, are frequently determined by several minor components, and thus are stable as long as a dominant component does not emerge. That is well known to mathematicians today, but it all had to be discovered, and for that to happen, quite a few outstanding minds had to consider the problem, from Abraham de Moivre, who discovered an early version of the central limit theorem in 1733, through Gauss, Galton, Pólya, and today's researchers who continually discover new variants of the central limit theorem and apply them to natural and social phenomena.

Perfect Symmetry from Perfect Asymmetry

Mathematicians have also generalized the central limit theorem in another direction. The symmetry of the bean machine—each ball, at each level, jumps with equal probability to the right or to the left—made it difficult for biologists to use this mathematical device as a model. In the world of living things, natural selection favors certain genes—those offering a selective advantage—over others. If we are to add natural selection to our model, we might say, for example, that a jump to the right offered a greater contribution to survival than a jump to the left.

Mathematicians have examined what happens if the board is biased. Suppose at each level, a ball lands on a small lever, which will tip to the left or right, and that these levers are all rotating to the left—counterclockwise, that is—like tiny propellers. As a result, a ball is always more likely to jump to the left than to the right. If the propellers are spinning very fast, the ball will almost always jump to the left, while if they are spinning more slowly, the probability of a leftward jump will be less.

With such a biased bean machine, the balls will no longer pile up symmetrically in the bins. More will land to the left and fewer to the right. Nonetheless, if the board is wide enough and tall enough, the balls will again approximate a Gaussian curve, only its peak will be somewhere to the left. The faster the propellers spin, the farther to the left it will be. The central limit theorem still works on our biased board.

The mathematics of the biased board shows another interesting property of the normal distribution. The Gaussian curve is perfectly symmetric, but its symmetry can arise from asymmetric components. In Mildovia, perfect asymmetry can lead to perfect symmetry, and it often does.

I am now going to jump ahead and show you a fractal in figure 8. We will study these strange objects in part III. It is not symmetric at all, yet it conveys a strong feeling of regularity. I have placed this image here because it illustrates a fundamental difference between Mildovia and Wildovia. In Mildovia, even total asymmetry can lead to perfect symmetry. In Wildovia, even a principle of total symmetry (called scale-invariance and self-similarity) can lead to asymmetry. Figure 8 shows a very deep form of regularity at work, something much more complex than simple symmetry.

Figure 8. A fractal.

5

The Extremities of Mildovia

If you're forever young at heart, you never look into anything in
depth.

We saw in the last chapter that profound asymmetry can lead to
a perfectly symmetric Gaussian curve, but there are some phenomena whose distributions are inherently asymmetric, and that's just
the way it is. For example, the distribution of household incomes
(figure 9) is asymmetric, since there is a hard limit of zero at the
bottom, but at the top, the sky's the limit. Because we do not expect
the perfect symmetry of the Gaussian curve, we do not expect the
distribution of incomes to be described by that curve with any great
accuracy. But because household income is determined by a number
of components, some version of the central limit theorem should
hold for incomes. Yet this appears to be contradicted not only by the
lack of symmetry. In addition, the right-hand side of the diagram
approaches the horizontal axis much more slowly than the Gaussian
curve, yet more rapidly than the Cauchy curve, and in fact it is closer
to Cauchy than to Gauss. Indeed, the last two columns suggest, correctly, that the long tail that begins in the second half of the figure
could be continued for a very long time. Extremely high incomes exist and are not even extremely rare.[1] Does this mean that household
income is a Wildovian phenomenon?

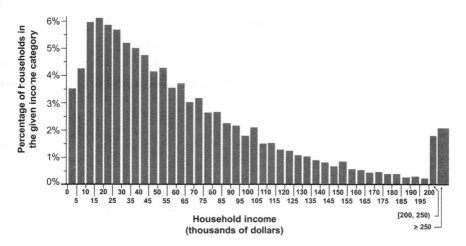

Figure 9. Distribution of household income in the USA in 2010 (drawing by József Bencze based on data from the U.S. Census Bureau website).

The Lognormal Distribution

Despite appearances, the Gaussian curve can be used to describe the distribution in figure 9. If we scale incomes logarithmically so that the distance along the *x*-axis from $1 to $10 is the same as from $10 to $100, and so on, the curve transforms itself into a nice, accurate Gaussian distribution. We call such distributions *logarithmically normal,* or *lognormal* for short, because they look normal when scaled logarithmically. Although truly large incomes exist, they do not belong to the world of Wildovia, because the Gaussian curve still models them fairly well. We have not yet left Mildovia.

With the aid of the logarithmic scale, mathematicians were able to come up with a new version of the central limit theorem.[2] If a trait is determined by several minor components and the components are not very strongly interdependent and there is a natural bound on one side (left or right) that does not allow for larger or smaller values, but there is no such limit in the other direction, then that trait will be distributed lognormally over the entire population.

Body weight is almost such a trait, but not quite. Weight cannot, of course, be negative, so it is naturally bounded, but it is also bounded from above, although the upper bound will be farther from the mean. Figure 10 shows the distribution of the weights of adult men in the United States in 2010. This graph is somewhere between the normal and the lognormal distributions. Weight also belongs to the world of Mildovia, even if some people weigh six hundred pounds. We should not consider such people miracles, just as people with millions of dollars of income are not miraculously wealthy. They are simply at the extremities of Mildovia. (This is generally true, but we shall see below that *extremely* large incomes should be considered a Wildovian phenomenon.)

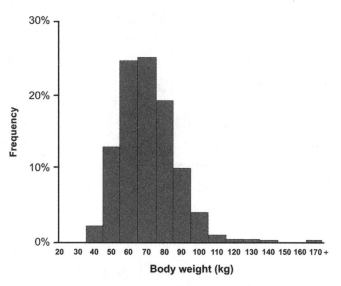

Figure 10. The distribution of weights of adult males in the United States in 2010 (drawing by József Bencze based on data from the Cancer Network website).

The Pareto Principle

The Italian sociologist and economist Vilfredo Pareto (1848–1923) also studied the distribution of incomes, and he was the first to apply the concept of social elite. Pareto was probably unfamiliar with the lognormal distribution, so he tried to devise his own model to describe incomes. He came up with a formula that approximated the observed distribution of large incomes relatively well, but it did less well with below-average incomes. This shortcoming did not appear to concern him. Pareto's formula has nothing to do with the formula for the lognormal distribution, though the curve it describes is fairly similar to it in the upper regions.

Mathematicians and economists still debate whether the Pareto distribution or the lognormal distribution better describes the actual distribution of high incomes.[3] They have also devised several other formulas that work well in different contexts. One thing, however, is certain. Although the formula for the Pareto distribution is even simpler than that of the lognormal distribution, it does not possess all those spectacular mathematical properties that we have discussed so far: no central limit theorem applies to it, and it does not guarantee any kind of stability. To a mathematician, the Pareto distribution looks like the work of a charlatan. Nevertheless, we must give Pareto credit, because his formula inadvertently anticipated the science of Wildovia. But before we can go much further in that direction, we first have to become acquainted with certain Wildovian phenomena. Even though mathematicians turned up their noses at the Pareto formula's lack of mathematical rigor and were uninterested in the real-world aspects of his work, Pareto had discovered an important rule about the distribution of incomes, though as it turned out, his conclusions can be proved more precisely in terms of the lognormal distribution than with the Pareto distribution.

Pareto observed that about 20 percent of all households take in about 80 percent of all household income. We can observe from figure 9 that in the United States, about 20 percent of households have incomes above $100,000, and with a bit more work, we can calculate that those households indeed receive about 80 percent of annual income.

Pareto found that this lopsided imbalance applied not only to incomes, but also to wealth and many other types of resources. In fact, his observation was found to be more or less applicable to a number of apparently completely unrelated phenomena. In most countries, around 80 percent of the population inhabits 20 percent of the settled terrain. Around 80 percent of the total mass of a galaxy is distributed among 20 percent of its stars. About 80 percent of oil fields are found on 20 percent of the Earth's surface. About 20 percent of forest fires consume 80 percent of all trees consumed in fires. The list goes on and on.[4] The idea that *around 80 percent of effects can be traced to only 20 percent of causes* is known as the *Pareto principle* or the 80–20 rule.

One reason the lognormal distribution (such as we saw in the distribution of incomes) does not belong to the world of Wildovia is that although it is not symmetric, it has a standard deviation. In the case of the lognormal distribution, if the natural bound at one end (such as the zero lower bound for income) is about two standard deviations from the mean, the 80–20 rule will apply.

The lognormal distribution thus not only provides a theoretical basis for the Pareto principle; it also tells us when the Pareto principle will be applicable. If a phenomenon is defined by several components, none of which is dominant, and there exists a natural lower or upper bound at about two standard deviations from the mean, while there is no natural bound in the other direction, then the Pareto principle applies. This sounds like a very specialized set of restrictions, but a surprisingly large range of phenomena fits within it. The concrete examples often cited in textbooks on business and management include the following:

- 20 percent of our activities earn 80 percent of our income.
- 80 percent of consumer complaints are caused by 20 percent of errors.
- we use 20 percent of the tools at our disposal 80 percent of the time.
- 80 percent of an athlete's performance is a consequence of 20 percent of his/her training.
- 20 percent of salespeople are responsible for 80 percent of sales.
- 80 percent of our cell phone calls are to 20 percent of the names in our phonebook.
- 80 percent of links on the Web point to 20 percent of all sites.
- 80 percent of music downloads are generated by 20 percent of music.
- 80 percent of a company's income comes from 20 percent of its clients.
- 20 percent of problems account for 80 percent of a company's losses.
- And a more frivolous example: 20 percent of the human population is having 80 percent of the sex. This statistic excludes prostitutes, because if we include them, we find that an even smaller percentage of the population is having an even larger share of the sex. Nevertheless, a more businesslike instance of the Pareto principle applies here: 20 percent of all prostitutes produce 80 percent of all income from the sex trade.

The accuracy of these examples depends on a number of factors, in particular the extent to which the conditions of the central limit theorem concerning the lognormal distribution apply. Are there several minor components? Are there really no dominant ones? Is there indeed a hard limit at one end? Is the sky truly the limit in the other direction? If all these hold, we still have to check whether the natural bound is around two standard deviations from the mean. If we can answer yes to all of these conditions, then we can expect the 80–20 rule to hold.

The Domain of Validity for the 80-20 Rule

Despite what "get rich quick" business courses and books repeat ad nauseam, the Pareto principle does not mean that this high-yield 20 percent is the only thing worthy of our attention. The multi-component aspect of the lognormal distribution tells us that even if we double our efforts in trying to get more out of the "important" 20 percent while neglecting the remaining 80 percent, not only will the result not double, it will actually decrease, because those components have already exerted their full effect. There is nothing more to be gained there. The way to achieve a surplus yield is to focus on the remaining components—the 80 percent that is underproducing. Nevertheless, it is worth becoming familiar with the Pareto principle, especially if we have to be selective as to where we put our energy.

The lognormal distribution also sharply points out that the Pareto principle is far from universal. If the natural bound is closer to the mean than two standard deviations, for example, then fewer than 20 percent of causes will account for more than 80 percent of effects. Book publishers are a good example. The natural lower bound is, of course, zero copies sold. A typical book sells one or two thousand copies in the United States, while sales of millions of copies are by no means rare. In this case, where the natural lower bound is closer than one standard deviation from the mean, the Pareto principle tells us that 20 percent of all published books account for not 80 percent but 97 percent of all sales.[5] Despite this fact, publishers continue to publish the remaining 80 percent, and they are quite happy with sales in the thousands for any single book. In addition to the Pareto principle, Chris Anderson's "long tail" also applies here: income is not increased by devoting more effort to the highest-yielding few percent but by focusing on the rest, if we have the will and energy to do so or if we're so small that we don't stand a chance in the first place of publishing a million-copy blockbuster.

If you want to apply the Pareto principle to your own business activities, first you need to estimate your typical yield and the typical deviation from it. Having done that, you can then check whether the natural lower bound is in fact two standard deviations from the average. If it is, the 80–20 rule applies. If it isn't, then completely different figures apply.

The 80–20 rule is often considered characteristic of Wildovia, but in fact it is a direct consequence of the normal distribution, so it belongs entirely to Mildovia. While this fact forcefully illustrates the power of Mildovian science, it raises a question: if so much belongs to Mildovia, what phenomena are not Mildovian? What requires a mathematical model that describes a completely different world? We will see in the next chapter that certain phenomena in the field of economics do in fact need completely different models. But first let us explore more of Mildovia's outer reaches.

Eternal Youth

Even the question of eternal youth is still within Mildovia. It is not only alchemists and Rosicrucians, Peter Pan, Dorian Gray, and

Juan Ponce de León who have something to say on the subject; so do mathematicians, and they speak about it in the language of Mildovia. They do not try to stay eternally young by remaining open-minded, nor are they especially interested in immortality. Instead, they concentrate on a more abstract, yet more mundane, sense of the term that is amenable to mathematical analysis.

If we state categorically that a person is young until he or she reaches a certain age, say thirty or forty, then that pretty much ends the discussion. Nobody can stay young forever. But chronological age is only one characteristic of youthfulness and perhaps not even the most important. We may also consider a young person as someone with a long life expectancy, and that is something else altogether. It is counterintuitive, yet true, that a child is "younger" at age one than at birth. A newborn's probability of reaching, say, age sixty is less than the probability of a one-year-old living another sixty years. In this sense, the one-year-old is younger. Later, the longer a person lives, the shorter the expected remaining life span. But perhaps not.

Mathematicians put the question in their own abstract way: is there a mathematical object that is not immortal, meaning it will die sometime with 100 percent probability, yet whose expected remaining life span is independent of how long it has been alive?[6] This question indicates that even an eternally young organism will die sooner or later. At a certain point, something breaks inside it and it dies, but the probability of such an occurrence is independent of how long the organism has lived up to that moment. The probability that our eternally young creature will live, say, ten more years is the same today as it will be (if the creature is still alive) one year from now, twenty-eight years from now, or any other number of years from now. This is not, of course, true of human beings. Right now, at age sixty-six, I have a much better chance of living ten more years than I will have at age ninety-four, twenty-eight years from now (if I am alive then).

The answer to the question is that such a mathematical object exists, and there is essentially one such function for each probability *p* of an eternally young creature living one more time unit. Mathematicians call this eternally young distribution the *exponential distribution*. We can see a graph of it in figure 11, where $p = 2/3$.

As shown in figure 11, after one time unit (let us say one year), the area under the curve between 1 and infinity on the *x*-axis (the shaded region) is 2/3, meaning that two-thirds of the population has a life span of more than one year and so is still alive after one year. Thus the probability of an individual having survived this first year is indeed two-thirds. Similarly, we can calculate that after two years, four-ninths of the population is still alive, meaning that two-thirds of the two-thirds of the original population alive after one year has

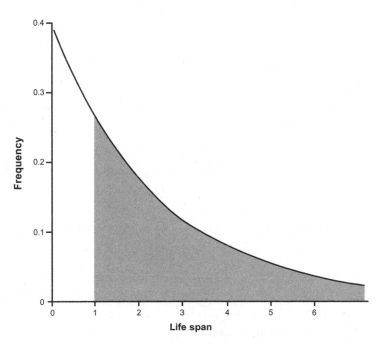

Figure 11. The exponential distribution (drawing by József Bencze).

survived one more year. And so on. At the end of each year, we find that two-thirds of the remaining population has survived one more year. (If the original population was A, then after t years, the surviving population will be $A \cdot (2/3)^t$.) The fact that the variable t is in the exponent is what gives this function the name *exponential*. The mathematical considerations that led to its discovery are in fact much less complex than those that led to the Gaussian curve, so this mathematical object is certainly also a product of the mathematics of Mildovia.

So it turns out that eternal youth, at least in this mathematical sense, is not theoretically impossible. But do such objects exist in the real world? This is no longer a mathematical question but a scientific one. Mathematics provided the idea that it is not absurd to search for such fountains of youth, and if one is found, then mathematics is ready to model it.

In fact, there are both natural and man-made objects to which this description applies. For example, a radioactive particle is bound to decay sooner or later, but the odds of its decaying in the next ten minutes are exactly the same as the odds of its decaying five years and ten minutes from now if it survives another five years. That is, the expected time until a radioactive particle decays is completely independent of how long it has existed. These particles are forever young, and their life expectancy conforms to the exponential distribution.

This description applies fairly well to some everyday objects, such as a neon light bulb. Sooner or later, every neon bulb will burn out, and usually it is sooner. But the expected time until it burns out doesn't depend on how long it has been burning. So a used neon bulb is, in theory, worth exactly as much as a brand new one. It is forever young.

The exponential distribution is also useful in describing phenomena such as the lengths of casual telephone conversations. For really dedicated gossipers, the length of a chat is exponentially distributed. This means that although the gossipers will hang up sooner or later, the amount of time they will continue to talk is almost completely independent of how long they have been chatting already. Gossip is forever young.

Eternal youth does not conflict with mortality—the two traits get along just fine—yet we do not find eternal youth in the biological domain. Organisms age and die. Why hasn't evolution come up with an eternally young plant or animal? You would think that immortality would be a selective advantage. (One possible candidate for immortality, or at least an approximation of it, is *Turritopsis dohrnii,* known as the "immortal jellyfish," which apparently can return to an early asexual stage after living as a mature adult. If you have ever wished you could live your life over again, you should consider turning yourself into one of these jellyfish.) The answer to this conundrum, to the extent that we presently understand it, is simple enough: the strategy that evolution has come up with for species survival uses genetic diversity to allow a population of individuals to achieve a certain stability in the face of environmental unpredictability. In other words, it is advantageous for the proportions of the various traits within the population to remain fairly constant generation after generation. If the environment changes—but not too much—the species will generally have enough variability to ensure that some individuals will survive, even if most die off. These survivors then reproduce and confer on their descendants a new mixture of traits, most likely better adapted to the new conditions. Such a strategy requires that old genetic combinations (parents) die off to make way for new ones (children), so it leaves no place for immortality.

Thus diversity and stability—two important underlying principles in biology—seem to exclude eternal youth. Perhaps not eternal life but definitely eternal youth. What is forever young is either unstable or uniform. One radioactive particle is just like another, and the same goes for neon bulbs. If you are forever young at heart—that is, open to everything throughout your entire life—you pay a heavy price: you never gain the wisdom of worldly experience.

Eternal youth is an extreme Mildovian phenomenon. Perhaps it is not such a big deal that it is incompatible with life and diversity. The exciting phenomena of Wildovia, in contrast, arise from other sources.

Zero-Probability Events

There is one more extreme Mildovian phenomenon that we must understand before we can venture into the science of Wildovia. I noted above that the science used in Wildovia is the same as the traditional science that describes Mildovia. What differs is the choice of scientific and mathematical models. For science is nothing but a set of methods accepted by the great majority of scientists. The general acceptability of methods is what gives science its guarantee of objectivity, in the sense that theoretically (and practically as well, for the most part), the validity of a scientific deduction or theory can be independently evaluated according to agreed-upon standards.

This is precisely what distinguishes science from other useful, and in many cases time-honored, methods of learning about the world, such as art, mysticism, philosophy, and religion. With a scientific result, not only do we know what we have just learned about the world, but we also know precisely how we reached the result. There is doubtless a great deal to be learned about the world from the *Odyssey* or *Harry Potter*. I myself have learned a lot from those books, but I can't accurately retrace the path by which I acquired that knowledge or even quantify just what I have learned. I can say only that those stories give an authentic account of the world around me, and reading them increased my knowledge of the world. But I can't even say precisely what made them so authentic, except that the authenticity is not a scientific one. Homer's *Odyssey* did not teach me that the classical gods intervene in human affairs, and *Harry Potter* did not teach me that there is a hidden realm of magic. What I learned has to do with human relationships, values, and passions.

Furthermore, because science works with models, we can understand the domain of validity of any particular result. Scientific truths are not absolute truths. They are valid only within the domain of the particular model, where results can be validated and refined by experiment. This is why the models used to describe Mildovia and Wildovia can be radically different yet valid each in its respective segment of the world.

I need to mention one more thing about the science of Mildovia if we are to understand the models describing Wildovia. The science of Mildovia can easily deal with the fact that just because the probability of a particular event is zero, that does not mean that it cannot occur. That may seem like a contradiction, but the logic behind it is quite simple.

Let us imagine we make a dot on a sheet of paper with a pencil at a random point. The probability that the center of that dot will be at any particular point on the paper is zero. The reason is that if the probability were any positive number, call it x, then x would be also be the same probability of any other point being hit. But there are infinitely many points on the paper, so the sum of all the probabilities over all the points of the paper would not be equal to one, as it must be, but infinity. Therefore, the probability of landing on a particular point is infinitesimally small—that is, zero.

But the pencil point must land somewhere on the paper, so when we touch our pencil to the paper, an event that a moment ago had zero probability has in fact occurred. The mathematics of Mildovia is able to handle this conundrum without any difficulty, although fig-uring out how probability would have to be defined to make a consis-tent theory required contributions from several outstanding talents, as well as a few geniuses. I will not, however, go into the technical details of how they worked this out.

If a zero-probability event occurs, that doesn't mean, according to the traditional science of Mildovia, that a miracle has occurred. Or we might say that it is in fact a minor, everyday miracle, what I have called a pseudomiracle. Thus the science of Mildovia can handle not only such low-probability phenomena as extremities of incomes or eternal youth. It can also handle events that occur despite their having zero probability of occurring. This is precisely why we keep bumping into pseudomiracles.

It is important to understand all of this clearly, because we will see that certain miracles in Wildovia are not characterized by a mi-nuscule or even zero probability but by not having a probability at all. That is, there are phenomena that cannot be assigned any probability, in the same sense that the Cauchy distribution has no standard devia-tion. That might be, for example, because they are inconceivable, but that is only one way in which it can happen. There are many other reasons why an event cannot have a probability assigned to it and can occur nonetheless, as was the case with our markswoman Phoebe, who could hit the wall almost unimaginably far away.

Gödel as an Extremity of Mildovia

In discussing the extremities of Mildovia, we shouldn't forget Gödel's theorem, which is entirely a product of Mildovian math-ematics, even if it shook that mathematics to its very foundations. It provided an extraordinarily extreme result that had been inconceivable to many of the foremost theorists. It was also the last thing a mathematician would have wanted. In 1900, when the great German mathematician David Hilbert compiled a list of the most urgent mathematical problems for the new century, near the top of the list was to prove that mathematics never leads to a contradiction.[7] This problem was solved by Gödel with his second incompleteness theorem, but the result was the opposite of what was expected. Instead of proving that mathematics was consistent, he proved that there was no possible proof of consistency within mathematics itself. It is said that Hilbert became angry when he learned of Gödel's theorem. John von Neumann, as we saw, simply gave up his research on logic.

Before Gödel, it was always taken for granted that finding a solution to a mathematical problem was only a matter of sufficient ability, intelligence, and cleverness. A mathematician was not 100 percent certain that a solution existed, because the problem of consistency set out by Hilbert had not been solved (and still has not), but there was no reason to doubt that mathematics would eventually be shown to be consistent. Since Gödel, one has always to bear in mind that the reason one has been unable to solve a problem might be not a lack of cleverness but that the problem is Gödelian—that is, that neither the proposition nor its negative can be deduced within the system of mathematics. And sure enough, results keep coming in proving that certain problems are, in fact, Gödelian.

If a mathematical assertion is Gödelian, then mathematics cannot determine whether it is true or false. A natural science, say physics or psychology, might be able to make a determination in a particular real-world context, but mathematics cannot give a definitive answer. All mathematics has to say is that as far as it is concerned, it would be just as happy with either an affirmative or a negative result. And indeed, our mathematical system can be extended with either the assertion or its negation as a new axiom! And mathematics will remain consistent in both cases (if indeed it is consistent), though each case may require a different type of mathematics. If, for example, physicists find that space has zero curvature, then mathematics will provide a model in the form of old-fashioned Euclidean geometry. But if physicists somehow discover space to be curved, then mathematics can offer a variety of non-Euclidean geometries, such as those cooked up by Bolyai and Lobachevsky. It is not a mathematical question whether space has zero curvature. In fact, physicists have come to believe that space has nonzero curvature.

As a consequence of Gödel's work, it has become clear to mathematicians that mathematics cannot exclude the possibility of several different systems of mathematics existing side by side, obeying different laws. We have seen, for instance, that civilizations may live side by side with very different concepts of integers. Some may have no concept of the number zero. Recall that the hyperreal, infinitesimally small number $1/I$ is also zero in a certain sense, in that it is smaller than every traditional real number. Yet it is not "really" zero, because we can divide by it, an action we could never perform with our traditional zero.

The fact that certain phenomena operate according to the laws of Mildovia does not preclude the laws of Wildovia applying to other phenomena. We have to construct the sciences of both worlds and consider carefully to which world any particular phenomenon belongs. This is how Gödelian thought, a first-rate product of Mildovia's science, provided the basis for understanding the miracles of Wildovia.

The change in attitude resulting from Gödel's discoveries made it possible to consider the phenomena of Wildovia to be just as valid as those of Mildovia. Although Mildovian science came up with the

Cauchy distribution, which seemed bizarre because it has no standard deviation, and came up with Gödel's theorem and then the idea of taking not-G as an axiom, which led to hyperreal numbers, the Mildovian mindset had no idea that such anomalous objects could apply to the real world. These mathematical phenomena seemed nothing more than theoretical objects of theoretical interest. We have yet to discuss what made the actual study of Wildovia necessary. The first field to need a Wildovian worldview turned out to be economics.

6

The Sources of Equilibrium

It is impossible to comb a hedgehog properly. There will always be
a spot with spines sticking out all over the place.

In 1910, the Dutch mathematician Luitzen Brouwer proved a
strange mathematical theorem.[1] Take a cup of coffee and stir it thoroughly, violently if you like, but in such a way that it remains a single
mass: no lifting a spoonful and pouring it back in. When you are
done stirring, let the coffee come to rest. Brouwer's theorem tells us
that there will be an atom in the coffee exactly where it was prior
to stirring. In other words, there is no way to stir a cup of coffee so
that every one of its atoms ends up in a different place from where
it started.

One might suppose that if such an atom existed, we could have
just nudged it a bit at the end; after all, nobody told us when to stop
stirring. But Brouwer's theorem guarantees that if we move that particular atom, another atom somewhere else in the cup will thereby be
moved back to its original location.

Of course, a mathematician doesn't build theorems around a cup
of coffee. (This is not to say that coffee plays no role in the production of mathematical theorems. The Hungarian mathematician
Alfréd Rényi, who was addicted to the stuff, once remarked that "a

mathematician is a device for turning coffee into theorems.") Mathematicians require precision, so they base their argument not on a cup of coffee but on a closed, compact, and convex set in some topological space; not on atoms but on points in that space; and not on stirring but on a mapping of the given set into itself. The constraint that the coffee be kept together is expressed by the constraint that the mapping be continuous. The theorem is then phrased as follows: A continuous mapping from a closed, compact, and convex set has a fixed point. This is called *Brouwer's fixed-point theorem.*

We can see how sloppy our coffee example is. It is no accident that a mathematician talks about much more abstract things than caffeinated beverages. A cup of coffee is not a closed set in the mathematical sense. Its boundary is the wall of the cup, which isn't part of the coffee, and as the coffee is being stirred, water molecules are evaporating from it. Nonetheless, the concrete example makes the theorem vivid; it provides motivation, even if it fails in accuracy.

If we remove the constraint of continuity, there is no longer the necessity of a fixed point. For example, if we somehow managed to separate the bottom half of the coffee from the top (say by freezing the coffee and slicing it in half) and then placed the top half on the bottom and the bottom half on the top, then every atom of coffee would be either higher or lower than it was before, and there would be no fixed point.

So Brouwer's fixed-point theorem guarantees that it is impossible, merely by stirring, to interchange the top and bottom halves of the coffee in their original configurations. We could keep on stirring this way and that, but Brouwer guarantees us a fixed point at every instant, and there is no fixed point in the configuration we are trying to achieve. Brouwer's fixed-point theorem is an important mathematical result in part because it is so counterintuitive; it really seems that there ought to be a way to stir every atom into a different location in the mug. In much the same way, it would seem that you ought to be able to comb a hedgehog. Here we are talking about a hedgehog that has rolled itself into a perfect ball. A corollary of Brouwer's fixed-point theorem guarantees that no matter how we comb it, there will always be at least one vortex, an area with spines sticking out all over the place.

Fixed-Point Theorems Worth Nobel Prizes

Mathematicians have generalized Brouwer's elegant theorem in many directions,[2] and a whole new subject, fixed-point theory, has emerged from the various applications of the theorem. It has been generalized to multidimensional spaces, to certain classes of discontinuous mappings, and in several other ways. The Japanese mathematician Shizuo Kakutani has even proved a fixed-point theorem for some abstract mappings that map a single point onto an entire set. Kakutani's generalization has since become one of the basic tools of mathematical economics. But bringing the real depths of Brouwer's theorem to the surface required a mathematician of John von Neumann's caliber.

Von Neumann didn't begin by generalizing the theorem; he set about applying it. His first application was something entirely unexpected: the development of game theory. Von Neumann discovered that what Brouwer called a fixed point could be understood in the context of strategic games—via a surprising and wide-ranging generalization—as a point of equilibrium. Here "equilibrium" means a set of strategies (one strategic plan for each player) whereby no individual player can achieve an improved position unilaterally merely through a change in strategy. (You can find a more detailed discussion of game theory in my book *Moral Calculations*.)

After he created game theory, von Neumann realized that if we view the subject of economics abstractly enough, it can be seen as a collection of games and mappings. Economics, then, consists precisely of two things to which Brouwer's fixed-point theorem applies.

The dynamics of game theory comes into play, for example, when any sort of deal is negotiated. The interests of the players are naturally just as opposed as the interests of poker players. But they may have common interests too, just as poker players have the common interest that the house charge be as low as possible and that there be no cheating.

Business transactions can be viewed as mappings. When we buy something, we map our money to goods. When we manufacture something, raw materials and labor are mapped to products. By construing such mappings abstractly, von Neumann discovered that an economy must contain equilibria, not only in the competitive, hence game-theoretic, aspects of transactions but also in areas as mundane as production and consumption.

It is highly significant that there exist equilibria in an economy, and in fact an entire economy can be in a state of equilibrium. In a way, it is surprising that economies are as stable as they are, given that every economic player is generally looking only after its own interests. In his great 1776 book *An Inquiry into the Nature and Causes of the Wealth of Nations,* the Scottish economist Adam Smith wrote, "It is not from the benevolence of the butcher, the brewer, or the baker that we expect our dinner, but from their regard to their own interest." He then went on to describe how self-interest can benefit society:

Every individual . . . neither intends to promote the public interest, nor knows how much he is promoting it. By preferring the support of domestic to that of foreign industry, he intends only his own security; and by directing that industry in such a manner as its produce may be of the greatest value, he intends only his own gain, and he is in this, as in many other cases, led by an invisible hand to promote an end which was no part of his intention. Nor is it always the worse for the society that it was no part of it. By pursuing his own interest he frequently promotes that of the society more effectually than when he really intends to promote it.[3]

Although the concept of economic equilibrium has since been defined much more precisely, the point is the same: the cumulative effect of individual selfish actions is to make the economy run efficiently. For a long time, however, Smith's ideas were accepted by neither politicians nor economists. The issue was too important, and Smith's examples, however spectacular, were too arbitrary. Can we really trust that the "invisible hand" will do the work previously attributed to private virtue and that it will look out for the public interest? Often it does not. Environmental degradation went on—and was ignored—for a long time, the invisible hand notwithstanding. It was not known where the invisible hand applied and what were the limits of its validity. In fairness to Smith, he did not claim that selfish interest *always* led to social good—in contrast to many of his more doctrinaire followers. For example, he warned against monopolistic practices and the undue influence of business interests in politics.

Nor did Smith precisely demonstrate the validity of the principle even in cases in which it actually works. His theory was entirely speculative. Much of *The Wealth of Nations* lists spectacular examples of well-meaning government interventions that turned out to be detrimental. We should therefore not be surprised that his book has repeatedly failed to convince socialist politicians of the wisdom of letting the invisible hand do its work, just as it fails to convince free-market politicians that at times the invisible hand requires governmental regulation.

Before von Neumann appeared on the scene, nobody knew how to convert Adam Smith's theory into a proper economic model whose domain of validity could then be examined. Von Neumann was the first to present economists with a useful model.

The 1994 Nobel Memorial Prize in Economic Sciences went to three researchers in game theory: John Nash (hero of the book and movie *A Beautiful Mind*), John Harsanyi, and Reinhard Selten. In 2005 it went to two more game theorists: Robert Aumann and Thomas Schelling. There are a dozen more Nobel laureates who were honored for an economic model based on Brouwer's fixed-point theorem or its application by von Neumann. Von Neumann himself was not awarded the prize, because he died in 1957, a dozen years before the first Nobel Prize in Economics was awarded. (It was not one of the original five prizes and is properly called the Nobel Memorial Prize.) One of the first was given in 1972 to Kenneth Arrow, who together with 1983 laureate Gérard Debreu, following von Neumann's ideas, proved in 1954 a mathematical theorem on the existence of general equilibria that is considered the fundamental theorem of equilibrium theories.

The Theorem of Arrow and Debreu

The Arrow-Debreu theorem (with some simplification) says the following:

If in an economy,

- there are no monopolies—that is, there is more or less a free market;
- no effects external to the economy appear (such as natural catastrophes, wars, or riots);
- business activities have no effects beyond the economy itself (such as pollution, corruption, or fraud);
- the economy behaves more or less continuously—that is, minor changes in the activities of individuals or companies result in minor changes in the total economy;
- supply and demand, including those of labor, are perfectly elastic;
- and a few more technical constraints apply, such as what economists call the law of diminishing marginal utility or diminishing marginal returns;

then there exists a state of equilibrium for the economy.[4]

This state of equilibrium is understood to mean that it is impossible to increase one's long-term gain by pursuing a strategy different from that implied by the state of equilibrium. Once such an equilibrium state appears, it can last for a long time, because it is in no one's interest to break it.

What guidance does this theorem provide to a government whose economy is more or less in equilibrium? We may assume that the government will try to maintain this seemingly ideal state of economic calm. To do so, it has only to ensure that the conditions of the Arrow-Debreu theorem are met, a goal it can accomplish by instituting policies that support those conditions even under altered circumstances. Such measures are necessary because even an economy in equilibrium is always changing—for instance, because of technological innovations, product improvements, population shifts, or weather- and climate-related phenomena. The job of the government, then, is to look out for violations of the conditions of the Arrow-Debreu theorem, such as monopoly formation or civil unrest, and to adjust policies accordingly. The rest can safely be left to the self-regulating force of the invisible hand.

At first glance, this state of affairs seems like a complete victory for Adam Smith. The invisible hand is able—with a little government help—to create a stable economic equilibrium, and this equilibrium is now guaranteed by the relentless rigor of mathematics. But all of it holds only as long the conditions of the Arrow-Debreu theorem are sustained.

Many of those conditions imply that the economy should be operating within the general limits of Mildovia. They do not, for example, allow for the existence of monopolies. While we remain in Mildovia, this constraint is reasonable, since monopolies are not created by a series of minor components but by a typically Wildovian mechanism, the so-called Matthew effect, which I will explain in a later chapter. Extreme extra-economic effects could easily lead to instabilities that are typical of Wildovia, so it makes sense to exclude them if we are looking for Mildovian calm. I have not discussed the meaning of such concepts as "continuity," "diminishing marginal utility," and "diminishing marginal returns," but they also more or less guarantee that an economy in line with the Arrow-Debreu theorem must necessarily be Mildovian.

It wasn't apparent from Brouwer's fixed-point theorem to what extent that theorem should be considered a product of Mildovian science. On the one hand, it had nothing to do with the Gaussian curve. Yet it had a mathematical constraint that hinted in that direction: Brouwer's fixed-point theorem applies only to closed sets. Hence the world must be closed, or anyway closed in a certain sense, for an equilibrium to exist. But the world of Wildovia is far from closed. In Wildovia, "miracles" are an everyday occurrence. In this light, it hardly matters which category of miracles they belong to: pseudomiracles; "true" miracles (phenomena unexplainable by the current state of science); or transcendent miracles, the sort of divine intervention that is outside the scope of this book.

Put Options

A look at options trading can give us a vivid sense of how an economy reflects both Mildovian and Wildovian influences.[5] Let us imagine that we are wheat farmers in a far-off country, where the unit of currency is the oru and grain is measured in korgels. We have sown our crop and calculated all our costs—including seed grain, the value of our labor from sowing until harvest, capital investment in farm machinery, repair costs, and so on. We have taken everything into consideration we needed to. According to our calculations, if nothing untoward happens and we get good weather, then we will recover our costs if we are able to sell our wheat at harvest time at 90 orus per korgel. This is good, since wheat is currently selling at 100 orus per korgel. We are nevertheless uneasy, for we have seen the price of wheat drop by up to 20 orus per korgel in a single growing season, and if our wheat brings us only 80 orus per korgel at the end of the summer, we're done for. Our farm will have operated at a loss, and we won't be able to buy seed grain for the next year. How much is it worth to us if someone promises now that they are willing to buy our wheat after the harvest at the current price of 100 orus per korgel? Would we pay 10 orus per korgel for such an option right now?

If at harvest time, the price of wheat has dropped below 100 orus, then we exercise the option. This is precisely why we bought the contract. Now we don't care how far the price of wheat falls. If is below 100 orus, we sell at the guaranteed price of 100 orus. With an option price of 10 orus, we have covered ourselves against the possibility of going bankrupt, because if we subtract the price of the option from the guaranteed price, we will be left with 90 orus per korgel, which is enough to cover our costs. If, on the other hand, the price rises above 100 orus, we would have no interest in selling our wheat for 100 orus since there would be buyers willing to pay more. After all, we have purchased an *option* to sell at 100 orus, not an agreement that we *will* sell at that price. In this case, we will be out the money we paid for the option to sell at 100 orus, but we are not downhearted about that because we have nonetheless made a tidy profit—how much will depend on the price of wheat at harvest time. But we will certainly net better than 90 orus per korgel. If the price skyrockets to 120 orus, we will be much better off. This "put option" is certainly worth the 10 orus per korgel we paid, because it insures us against bankruptcy if things go badly while leaving us the opportunity of a profit should things turn out better.

Call Options

Now let's imagine that we are not farmers but millers. We do not sell wheat; we buy it, mill it, and sell the flour. Although the price of flour is related to the price of wheat, the two prices do not fluctuate in sync; rather, there is a lag time. So our interests are opposed to those of the farmer: the cheaper the wheat, the happier we are, and if the price of wheat goes up, our profits are squeezed.

Like our friend the wheat farmer, we have also done the calculations to determine our operating costs, including the capital investment and the occasional necessary repairs to our mill. We have found that as long as the price of wheat does not exceed 110 orus per korgel, we will do all right. At 110 orus per korgel, we just break even. Like the farmer, we are pleased that the current price is 100 orus, for if we can buy at that price, we can expect a decent profit. But if it should rise to 120 orus at harvest time, we're done for. We might as well close down our mill, since we're just going to lose money on every korgel of wheat we process. Like the farmer, therefore, we need an insurance policy. How much are we willing to pay for a guarantee that we can buy wheat at harvest time at the current price of 100 orus? Let us suppose that just as for the farmers, this option is worth 10 orus per korgel to us millers.

The farmer needed a put option to insure against going bankrupt, while the miller needs a "call option" for the same reason. The farmer and miller could, of course, simply have made an agreement between them that next summer the farmer would sell his wheat to the miller for 100 orus per korgel (i.e., at the current price), and both would get something out of the deal. But each can do better: the farmer will make a killing if the price of wheat soars, while the miller will do the same if it plummets. Each would like to retain the possibility of making a fat profit, so instead of buying and selling in advance at the guaranteed price of 100 orus, both farmer and miller are willing to pay an option fee of 10 orus for the possibility of large gains without the risk of going broke.

An entrepreneur is used to having good years and bad years. In worse years, the aim is to survive, and options make survival possible. In better years, the aim is to make as much profit as possible. For the farmer, the good years happen when the price of wheat is unusually high, while for the miller, they happen when the price is unusually low. Still, they are not yet out of the woods. There are certain risks that concern them both—for instance, if the population loses its taste for bread and everyone starts eating rice. If that were to happen, a price of 100 orus per korgel would not be particularly profitable for either of them. They would be better off trying to find an option that wasn't too expensive.

The Investor's Motives

Now let us imagine we are neither farmers nor millers but investors. Our job isn't to produce wheat or to process it but to find suitable financial vehicles in which to invest. We understand the needs of the farmer and the miller, so suppose we want to offer one or both of them an option. As prudent investors, we need to determine how much we should charge so that in the long run we make a reasonable profit. The farmer and the miller are interested primarily in the short term: they don't want this year's price to drive them into bankruptcy. Investors, on the other hand, don't mind a short-term loss, provided there is sufficient capital available to cover such losses for a significant period of time. Investors want to ensure that their portfolio of investments will be profitable in the long run. If they can offer an option at a price that the farmer or the miller finds reasonable, then everyone gets something: the farmer and the miller are insured against catastrophic loss while retaining the opportunity to win big, and the investor will gain in the long run.

If we offer options to both the farmer and the miller at 10 orus per korgel, both of them will find that price reasonable. But we are competing with other investors, and perhaps another investor will be content with an option price of 9 orus per korgel. For an investor who can afford to take a very long view, it might make sense to offer options at a cut rate in order to acquire a large client base. Of course, the danger is that a miscalculation could lead to eventual bankruptcy—which could happen even with a 10-oru option price.

For us as investors, the best years are those in which the price of wheat is right around 100 orus, because we have a guaranteed 10 orus per korgel from the farmer, the same from the miller, and yet no loss from having to buy or sell at a disadvantageous price. If the price of wheat goes down sharply, we have to buy the farmer's low-valued wheat at a high price. If the price goes up sharply, we have to sell expensive wheat to the miller for less than the market price. In the latter case, we can hedge our bets by buying wheat futures at the current price—that is, we agree to buy a certain amount of wheat at harvest time at today's price. Let's suppose that we purchase a futures contract for one korgel of wheat for each two korgels' worth of options sold. Of course, if the price drops, buying wheat futures won't have been a profitable move. But the loss there will be partly compensated by income from the option sale. This is why a call option is always cheaper than a put option.

We see, then, that selling options involves risk, and when investors sell options, they usually take other measures to reduce their risk, such as mitigating the risk of a put option by buying some wheat futures at the current price.[6] Such an idea would never cross the farmer's or miller's mind. Why would the farmer want to buy from himself the very wheat he is producing? This is why both farmer and miller are better off not doing business with each other but with the investor.

This is a very simplified description of how options work. In fact, the price of a call option is generally not the same as that of a put option. All things being equal, the price of a put option could be many times that of a call option. If the farmer and the miller had reached an agreement with each other, it would not have been a fair deal. Despite all appearances, the farmer would have been the winner, since his option is the more expensive one.

I've simplified the description in another way as well, by discussing only call and put options. In reality there are many other kinds of options, but my goal is not to write a textbook on options trading. I brought up the subject only because it will soon become apparent that options demand Wildovian economics.

The Pricing of Options

Fischer Black (1938–1995) was a physicist who received his doctorate in mathematics from Harvard University. Nowadays it is fairly common for someone with such a background to become a financial researcher, but in the 1970s, when Black did it, it was exceptional. He became interested in the problem of how to set—at least theoretically— the correct price of an option.

Back then, options contracts were made pretty much by the seat of the pants: if a price was agreeable to both the seller of the option (the investor) and the buyer (the farmer or miller), then the deal was closed. If they failed to reach an agreement, there was no deal. Many farmers and millers went bankrupt under this arrangement. Black wanted to find an objective way to measure an option's actual value. He saw clearly that an investor's life is far from simple. Exchange rates fluctuate continuously. And everyone's sensitive point is at a different price: one farmer might be willing to purchase an option to sell at 100 orus per korgel, while another one might have no problem with an option for selling at 95 orus, though he won't pay as much for it. A farmer might want an option for a year in advance, while a miller might settle for six months. The owner of a technology firm might need an option for only a month. Black wrote down all the equations that resulted from such considerations, but they were too complicated for him to solve. Then he met Myron Scholes.

Scholes, an economist, discovered that the theoretically correct price of an option is independent of the actual future price of what is being optioned. All that counts is how much the commodity's price fluctuates, or in other words, the standard deviation (economists call it "volatility"). This was the idea that required a professional economist, someone who was familiar with such things as the existence of equilibria and the Arrow-Debreu theorem. Scholes added a few points to Black's equations (such as the conditions of the Arrow-Debreu theorem), making it much more likely that they could be solved. But the equations were still too complicated.

It was Robert C. Merton who complemented Black's and Scholes's theoretical considerations with his knowledge of investors' habits. Black's equations were overly complicated because he had considered every theoretical possibility, while real flesh-and-blood investors don't care, or need to care, about every possible situation. An investor is not a theoretical expert. He doesn't compute integrals; he does business. And when he sells an option, he automatically, almost as a reflex, takes some practical measures that will reduce its risk. Once Merton reduced the number of cases that had to be considered, the solution to the equations became expressible as a single mathematical formula.

Thus in 1970, the *Black-Scholes formula* was born.[7] It became one of the most widely used formulas by investors in the last quarter of the twentieth century. It did not appear in print until 1973, after it had been rejected by every leading journal because the editors found the mathematics too convoluted. Myron Scholes and Robert Merton later received the Nobel Prize in Economics. Fischer Black didn't live to see it. It is rumored that the reason Robert Merton wasn't commemorated in the formula's name is that on the morning Black and Scholes first presented their results at a conference, he overslept.[8]

The Black-Scholes formula makes it possible to evaluate the theoretical price of any option. The formula is complicated, but the calculations are done by a computer. (Many free Black-Scholes calculators can be found online.)[9] The fact that the formula is complicated does not interfere with the ease with which it can be applied. And it *is* easy. All you have to do is input the current price, the strike price (the price at which the investor is willing to guarantee to buy or sell), whether it is a call or a put option, and the expiration date (the date on which the option can be exercised). You must also give the volatility (standard deviation) of the underlying asset, as well as the currently largest available essentially risk-free rate of interest, such as the rate of interest on deposits at the most stable banks or the yield of U.S. government bonds. The Black-Scholes formula will return a single number: the theoretical price of the option.

The Case of the Investor Walking By:
Marina at the Marina

After they had come up with their formula, Black, Scholes, and Merton wanted to know how real-world option prices compared with those proposed by their formula. They found a surprisingly large variation. Compared to their formula, some options were considerably underpriced while others were wildly overpriced. The most underpriced option they found was a call option for shares of a company called National General. Black wrote years later: "Scholes, Merton and I and others jumped right in and bought a bunch of these warrants. For a while, it looked as if we had done just the right thing. Then a company called American Financial announced a tender offer for National General shares. The original terms of the tender offer had the effect of sharply reducing the value of the warrants."[10]

The three scientists lost their investment. American Financial paid a higher price for National General shares than the strike price, so the scientists' option became worthless. On the one hand, they were unhappy with their financial loss. But they were pleased that their formula, at least for a while, had correctly predicted the value of the option. It wasn't their mathematics that had failed them; it was their business acumen. Call options for National General shares had been underpriced precisely because rumors of such a buyout had long been in the air, and the Wall Street pros had been counting on it. Black concluded: "The market knew something that our formula didn't know. . . . Although our trading didn't turn out very well, this event helped validate the formula. The market price was out of line for a very good reason."[11]

Why did American Financial find it worthwhile to pay a much higher price for the shares than they were "actually" worth? We can explain the company's logic through a very simple example. Let us imagine that we have a small fishing enterprise. We have a fishing boat, and by casting our nets out at sea, we are able to catch more fish than the boatless fishermen who have to fish from the beach. But our company is not doing very well. For some reason, it is not living up to expectations, so it is valued low by the market. But an investor walking by—let us give her the seafaring name Marina—notices that when the captain isn't looking, the crew is lounging on deck in the fishing nets, soaking up the sun. It is no wonder the enterprise isn't doing very well. The problem is not something Marina can readily change, but it occurs to her that perhaps this boat shouldn't be used for fishing at all. With a little refurbishing, it could, for example, be used for pleasure cruises, and then it would make much more money than it had in fishing.

Marina could just build herself a cruiser from scratch, but that would take time, and in any case, the whole reason people take boat cruises in this little fishing village is that they want to put out to sea in a traditional fishing boat. The cruises could even allow passengers to catch their own dinners. An old, weather-beaten boat would provide just the right nostalgic touch. But the boat would have to be stable, and it would have to be brought up to code. So Marina decides that she doesn't want a brand new vessel. It would be a better investment for her to buy and renovate the fishing boat, and she would be willing, if necessary, to pay somewhat more than the price at which the market currently values it.

Marina keeps quiet about her idea until she reaches an agreement with us, the owners of the fishing company. She might even keep her idea to herself as long as possible so that imitators are kept out of the market. In this respect, our example differs from the case of National General, where American Financial's intent to acquire the company couldn't remain a secret, and the market wisely put a price on the transaction before it happened. The market may not have understood American Financial's motivation, but it did sense that something was afoot.

It is this "something in the air" that the Black-Scholes formula does not account for. Nor can it, since "something in the air" is not a mathematical concept but a bona fide business idea. Marina walking by the marina showed more signs of genius than the Black-Scholes formula, because she was the only one to think of something that had never crossed anyone else's mind.

Although the Black-Scholes formula certainly deserved a Nobel Prize, every element of it had been in the air since the 1960s. It took three experts from different fields—mathematics, economics, and business—to pool their backgrounds and intellects to solve all the technical problems. And they were aided by many other researchers.

The Black-Scholes formula, then, may be seen as the result of hard work by talented, seasoned professionals, but it is certainly not a work of genius.

For the formula to be valid, the fluctuations in price of the underlying asset (for whose sale or purchase an option is offered) need to have a Mildovian distribution. If prices were to follow, say, the Cauchy distribution, it would become impossible to quantify the fluctuations in terms of a standard deviation, and we could not even begin to use the Black-Scholes calculator. But the mathematics underlying the Black-Scholes formula is Mildovian in a more obvious and unequivocal sense. The formula explicitly supposes that the distribution of price fluctuations is described by the Gaussian curve. And it is described by that curve as long as the economy is hanging out in Mildovia and price fluctuation is determined by a multitude of minor components. An idea like that of our investor friend Marina, however, is a single big idea. Such ideas, which are not rare, have made many people wealthy. And they lead the economy inevitably into Wildovia.

The Rich Man's Junk Heap

The Polish-born French-American mathematician Benoit Mandelbrot (1924–2010) will be the hero of part III of this book, the part about the science of Wildovia. He has, in fact, already appeared, for it was he who discovered how to create fractal images like the one shown in figure 8.

In the 1960s, when the mathematical apparatus that would lead to the Black-Scholes formula was just evolving, Mandelbrot argued at several conferences that the Gaussian curve might not describe accurately the actual behavior of prices. He proposed using the Cauchy curve as the foundation of such models. In other words, Mandelbrot suggested back in the 1960s that economic models be moved from Mildovia to Wildovia.[12] This suggestion drew little enthusiasm from economists. Mandelbrot's not-exactly-smooth manner may have played a role, but that wasn't the main point. If his suggestion had been taken seriously, economists would have had to trash everything they had come up with since von Neumann, Arrow, and Debreu and exchange the reassuring models of equilibria for something a lot less elegant.[13]

If the economists had done that, Black et al. might well have concluded than an option cannot have a theoretically correct price. You cannot confidently put a price on something with no standard deviation. And even if you could—for example, by domesticating the wild Cauchy distribution in the model—it would lead to option prices that were so high that no one would want to buy them. Our farmer and miller would be back at square one. But practical experience had already shown that options can be priced in a way that both parties will consider reasonable.

The price fluctuations that would have justified such a radical change to the models had not been seen since the Great Depression of 1929. That was so until 1987, when a huge economic crisis shook the world, and again in 2008, when we suffered an even bigger one.

Everyday life still goes on in Mildovia, even if Wildovian phenomena ruin the picture from time to time. After all, options contracts can be drawn up only under Mildovian conditions, since in Wildovia, options simply don't have a theoretically correct price. And the farmer and the miller need such options as a hedge against bankruptcy. The investor is happy, too, as long as Mildovian rules apply. In the long run, in fact, the investor does even better than the farmer or the miller. When the chaotic laws of Wildovia finally come home to roost and a negative black swan appears on the horizon, it does not devastate the farmer or the miller. It will be the investor who loses his shirt. The farmer's and the miller's options payments have allowed the investor to do well as long as the rules of Mildovia prevail.

The moral is this: if you want to make your living as an investor, you had better be able to withstand a total crash and pick yourself up, dust yourself off, and start all over again. You are not a "real" investor if you haven't gone bankrupt a few times and each time gotten back on your feet. My friend Alex has hit rock bottom a few times in his life, but he has always been able to pull himself up and become even more successful than before. Aside from his mantra "There are no miracles," the line I have heard most often from Alex is, "There is always the next opportunity." But for an investor to make good use of the next opportunities, he has had to succeed with the previous opportunities, even those that eventually led to bankruptcy. For an investor, bankruptcy isn't a sign of failure; it is just an uncomfortable but necessarily recurring side effect of the job.

It is said that there is more value to be found in a rich man's junk heap than in all of a poor man's possessions. In a way, the investor's junk heap is what the farmer and the miller finance through their options fees, and its existence is what gives an investor a better chance of recovering from bankruptcy than a farmer or miller has, although it is also true that an investor needs the sort of personality that can accept the profession's inherent volatility. All of this foreshadows the question at the heart of part IV of our book: how can we achieve some kind of balance between the equilibrium-based world of Mildovia and the disequilibrium of Wildovia? First, let us pay a visit to the rough-and-tumble world of Wildovia.

PART THREE
The Wild World

The idea isn't not to lose in case of a crisis;
it's to make use of the flip side of the coin,
positive miracles, when there is no crisis.

7

The Mathematics of the Unpredictable

It is the brain that works chaotically, not the heart.

The Palace of Miracles in Budapest has on display, along with many other fascinating exhibits, the mechanism shown in figure 12. Three arms, each consisting of two hinged segments, extend from the center of a wheel. Each arm is therefore a double pendulum — that is, a pendulum with another pendulum attached to its end. The three pendulums are anchored by a screw at the center of the wheel. If you choose a starting position for the three double pendulums and then release them, they start swinging. Now comes the big surprise: the outer arms of all three double pendulums appear to be moving in a completely random way. There is no apparent rhyme or reason in their motion. Sometimes one of them will swing upward and might even, when you least expect it, flip over like a pendulum ride at an amusement park. Start the pendulums again. And again. Each time, the result is completely different.

Most visitors stand mesmerized in front of this display. Some might actually lapse into a state of mild hypnosis after staring at the three arms moving in such a spectacularly unpredictable way.[1] More than the chaos, it is the great variety of motions that makes this exhibit a suitable vehicle for us to focus on. According to current research on

Figure 12. Double pendulum at the Palace of Miracles, Budapest
(photograph by Péter Tamás).

hypnosis, it is precisely this state of focusing that is the only necessary condition for entering a hypnotic state. Everything else—relax-
ation, the hypnotist's words, even the presence of the hypnotist—is
dispensable.

My daughter was eight years old when she spontaneously discovered how to enter a trancelike state. When her brother, who is five
years older, asked her how she did it, she answered enthusiastically,
"Concentrate with all your might on something, perhaps a candle
flame or even just a dot on the wall." And when her brother asked,
"Should you close your eyes?" she replied, "That isn't important.
Although for beginners, it may be better to close them."

The Unpredictable Pendulum

The double-pendulum contraption may be an excellent tool for meditation, but our interest is scientific: is the motion of the three pendulums truly chaotic, or does it follow a pattern that we are simply not smart enough or observant enough to discover? Intuition favors the latter answer. It seems improbable that such a simple device, all of whose motions are completely determined by the spin of the wheel, could lead to completely unpredictable behavior.

Our intuition is wrong. In recent years, mathematicians have proved that even the motion of such a simple device as a double pendulum can be completely unpredictable. We don't even need three double pendulums; one is enough. Figure 13 shows the trajectory of

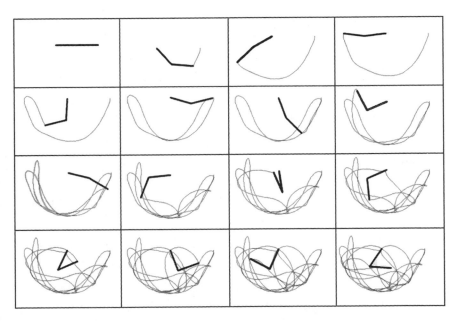

Figure 13. The chaotic trajectory of a simple double pendulum (https://en.wikipedia.org/wiki/Double_pendulum).

such a double pendulum with both sections starting from a horizontal position. More precisely, the figure shows the trajectory after one such start. The next time, even if it starts in almost exactly the same way, the trajectory will be not even remotely similar.

The evolution of the double pendulum's trajectory is a good example of what has been called the butterfly effect. Since it was coined in the 1970s, the expression has entered popular culture, serving as the title of the 2004 psychological thriller written and directed by Eric Bress and J. Mackye Gruber. It is also the name of a poetry collection, a murder mystery, and a rock band.

The Butterfly Effect

The history of the phrase "the butterfly effect" begins with the American meteorologist Edward Lorenz, who wrote a paper in 1963 based on a puzzling occurrence.[2] He was using a computer to run a weather simulation, and when he ran it a second time, he was feeling a bit lazy, so he typed in 0.506 for one of the parameters instead of the full value 0.506127. That minuscule change resulted in an entirely different weather pattern. How could that be? There was nothing random in the simulation; Lorenz had used a perfectly deterministic model. The only difference was that one number had been changed by one-tenth of 1 percent. He was astounded at how radically this small change in initial conditions could affect the outcome.

Lorenz investigated this phenomenon thoroughly, and out of his researches evolved chaos theory, certain aspects of which had been described half a century earlier by the great French mathematician Henri Poincaré (1854–1912), who, however, had failed to come up with such a catchy name—a big mistake, because in the fame sweepstakes, merchandising counts for a lot.[3] Thus we do not associate the invention of chaos theory with the name Poincaré. Not that Poincaré's shade needs to feel neglected; there are dozens of mathematical and physical concepts connected to the name Poincaré, though none of them has achieved the popular appeal of chaos theory.

Lorenz concluded that certain mathematical systems (such as his model for weather forecasting) operate in such a way that minute changes in the initial data lead to enormous changes in the system's behavior. Previously, mathematicians had believed that the hallmark of a well-behaved deterministic system was that small perturbations in the initial data (such as minor errors in measurement or rounding errors) would lead to only small effects on the outcome.

In his first paper on the subject, Lorenz wrote that a meteorologist colleague of his had commented, "If your theory is correct, one flap of a seagull's wings would be enough to alter the course of the weather forever." How the seagull transformed into a butterfly is lost in the mists of folklore. Some say an editor with a keen sense of style made the change. Others maintain that Lorenz himself referred to a butterfly in a later lecture. Anyhow, "butterfly effect" entered the vernacular, and one can hear in polite conversation that a single flap of a butterfly's wing in Brazil can cause a tornado in Texas. Sometimes the butterfly is in Tokyo, and the result is a hurricane in New York.

The danger of a brilliant metaphor is that it can confuse the true nature of the underlying object. Butterflies are no doubt flapping their wings all over Tokyo, and even more in the Brazilian rain forest. But despite all those wing flaps, we know, as Eliza Doolittle noted, "In Hertford, Hereford, and Hampshire, hurricanes hardly happen." The butterfly in Tokyo did not *cause* the hurricane. Rather, a small variation in wind currents between two weather patterns can result in an enormous variation in meteorological outcomes, so the rain in Spain ends up in the mountains instead of in the plain.

In fact, chaos theory provides a few elegant mathematical models in which the effect of a slight change in initial conditions from one iteration to the next is amplified and reamplified, leading to vastly different results, even if the initial difference is in the tenth decimal place in only one measurement. In such cases, the change in an air current caused by the beating of a butterfly's wing could conceivably be the first tiny step in a chain of events that leads to a tropical storm. But that does not mean that the butterfly caused the storm. It was caused by the nature of weather, which sometimes conforms to the model of chaos theory, which tells us that a large effect can be the consequence of a tiny cause.

Chaos theory has made us aware of an uncomfortable characteristic of our world: there are phenomena we cannot predict, and the reason is not our ignorance or inability to make precise measurements. Rather, unpredictability lies in the very essence of such phenomena, and when we encounter such a beast, neither more knowledge nor more precise measurement will help us, for both the real-world phenomena and the mathematical constructs we use to model them are inherently chaotic.

And these phenomena are not merely theoretical curiosities. They arise in the real world, and they are not rare, and what is most surprising is that some chaotic systems are much, much simpler than the weather. The double pendulum is an example. A double pendulum like the one in the museum in Budapest is infinitely simpler than Earth's atmosphere. Nevertheless, we can set its initial state only to a certain precision, and even a few thousandths of a millimeter of difference in the starting points of two releases of the pendulum will result in a radical change in trajectory. The model is completely deterministic. There are no hidden random effects. Yet the behavior is chaotic.

The rise of computing power has made it possible to model the weather with greater and greater precision. Yet the mathematical models that have emerged look more and more like chaotic dynamical systems. If aspects of weather as a natural phenomenon conform to the mathematical models of chaos theory, we will never be able to model weather so precisely that we can make accurate predictions all the time. In fact, the situation is even worse: the number of major forecasting errors cannot be improved beyond a certain limit. The number of minor errors, on the other hand, can be continually reduced, because those errors arise from nonchaotic aspects of the weather in which small deviations in today's atmospheric conditions will not be amplified into large deviations in tomorrow's weather. But the weather's chaotic components guarantee that no matter how much the science of meteorology advances, some enormous forecasting blunders will always happen.

From a mathematical point of view, however, things are a little better. Although there is no way to calculate the exact state of a chaotic system, it is nonetheless possible to compute the probability of a particular state occurring within a given time frame. Unfortunately, this is of little practical help in the case of a chaotic system, because the probability of any given extreme state is negligible, and you wouldn't want to issue a tornado warning for an event that has a one in a million chance of occurring. A weather forecaster who cried wolf too many times would soon discover that no one believed any of his forecasts.

Almost anything can happen in a chaotic system. If our global weather contains subsystems that exhibit behavior modeled by chaos theory, then the farther into the future we try to make predictions, the more likely it is that we will stumble across such a subsystem. Thus there will always be unpredictable meteorological phenomena:

a bolt can always come out of the blue. Chaos theory has shown us that weather is unpredictable in the long run, so although Earth's mean surface temperature has risen a significant amount in the last century, we cannot be absolutely certain that the warming will continue or whether perhaps a new ice age will intervene first. We can put chaos theory to better use by studying various possible scenarios and trying to quantify their likelihoods of occurring.

When we happen to be far from any chaotic state, we can predict with reasonable accuracy whether, for example, there will be a warm or a cold spell, a storm or sunshine. But when the system becomes chaotic, all bets are off.

It is entirely possible that chaos theory offers a good model of some underlying principles of nature, although we do not yet know how to formulate those principles or their domain of validity. In the next chapter I will describe a general principle that may form the theoretical basis of certain kinds of chaotic phenomena, but let us first see the multitude of areas to which various models of chaos theory can be applied.

Chaos in the Brain and Heart

We see in nature that some systems operate according to the Newtonian worldview, in which small changes in causes lead to only small changes in effects. But there are also systems whose rules of the road are better described by chaos theory. This is what I mean when I say that the laws of Mildovia apply to certain natural phenomena, while the laws of Wildovia apply to others. (There may also be phenomena to which neither set of laws applies, but that is not my subject here.) To see such systems at work, we have only to look within ourselves.

Recent research has suggested that the way the human unconscious mind works may conform more or less to models of chaos theory.[4] I find it amusing when one of my psychology students attempts to analyze his own unconscious, showing a budding competence while also revealing that some concepts are only half understood. "I've just made an unconscious slip of the tongue." "I have just discovered that unconsciously, I want to. . . ." These students have forgotten that the reason that we call it the unconscious mind is that *we aren't conscious of it*. And if the unconscious mind is in fact something to which chaos theory applies, then we can never be fully conscious of it.

The healthy human heart is basically Mildovian, and a pathological condition may be revealed by a chaotic element in an electrocardiogram. The healthy brain, on the other hand, is flagrantly Wildovian. Electroencephalograms, which show the electrical activity of the brain, have shown that one component of brain waves can be modeled with very good precision by a chaotic dynamical system.[5] These waves are "regularly chaotic," meaning that they are completely irregular on a regular basis. In certain illnesses, however, these fluctuations disappear to such a degree that their absence can be used as a diagnostic indicator—for example, in predicting a tendency toward epilepsy.

The heart and the brain have such fundamentally different operating modes that they have become symbols of two opposing attitudes. We consider emotions to be chaotic and dictated by the heart, and we believe our brains work in a stable, predictable way. Chaos theory suggests that our views about these two organs should be reversed. Even though the Bible tells us to "apply thine heart to understanding," we know today that thinking doesn't take place in the heart at all. But it took a long time for science to recognize this. This question of head versus heart is what the last major experiment of the great French chemist Antoine Laurent de Lavoisier was all about.

Lavoisier was not only a scientist, but also a nobleman, a status that was not good for his head during the French Revolution. In 1794, he was tried, convicted, and sent to the guillotine as one of a group of twenty-eight *"fermiers généraux,"* or tax farmers, who had become wealthy as collectors of revenue for the king. On learning his sentence, he realized that he had been given the chance of a lifetime, for he could finally decide one of the great unanswered questions in physiology: where is the home of the will, in the brain or in the heart? We should refrain from viewing this question with the smugness of our current knowledge, for today we know that the brain does the thinking, while the heart pumps blood. But back then, there were many signs suggesting that the heart might be an agent of will. For instance, if we really want something, we never feel anything in our brain, while we might feel our hearts swelling with longing.

Lavoisier devised the following experiment: he would blink his eyes for as long as he could after his head had been severed from his body. If the head could blink without the mediation of the heart, it would prove that the seat of the will is in the brain. The longer he could keep blinking, the stronger would be the evidence that the will to do so came only from his head. He instructed his assistant to take thorough notes and to publish the results under Lavoisier's name. He went under the guillotine happy in the knowledge that he had one more important mission to perform for science. And sure enough, when his head was severed, Lavoisier blinked continuously for fifteen seconds. Other sources say twenty seconds; others, as long as thirty.

I have read this story in many places. It is almost certain that not a word of it is true. It is an example of what would later be called an urban legend. There is no evidence of any publication by the assistant. None of the eyewitness accounts of the execution mention the blinking. There is no trace in university syllabuses that an account of such an experiment has ever been taught. We certainly do not teach it today, although we now have much more compelling (and less gruesome) evidence to support the contention that the will, like thinking generally, is seated in the brain. But even if this wasn't how it happened, the story shows the long road to the discovery that our ability to reason logically is connected with a chaotic component of brain activity that seems to exist only in *Homo sapiens*. Nonetheless, our memory is relatively trustworthy, our personalities change slowly and only seldom dramatically, and we are able to think more or less coherently. How do we reconcile the stability required to do this with the chaos in our brains?

The Simplicity of Chaos

Chaos, in the sense that mathematicians and physicists use the word, in no way excludes the possibility of stability. In fact, it even guarantees certain kinds of stability, although not in the sense that is familiar from everyday life.

Mathematicians and physicists consider a system chaotic if the following three properties apply to it:

1. A small number of variables (five to ten) determines the state of the system and determines it in a very simple way.
2. The system is very sensitive to small differences in the initial state.
3. At some time in its evolution, a system finds itself arbitrarily close to every state it is theoretically capable of reaching, even though it will not necessarily reach every possible state.

Like all mathematical statements, these three constraints are stated here too concisely to be understood without further elaboration. Let us consider them a bit more closely so that we can understand what chaos is and what it is not.

The first condition reflects the observation that even very simple equations can have extremely complex solutions. The double pendulum is an example; its motion can be described by three simple equations, and yet the pendulum can take a very complicated trajectory. The point of mathematical chaos is precisely that it can be brought about by very simple, even wholly deterministic, conditions.

The second condition is another way of stating the butterfly effect. It is a characteristic property of chaotic systems that minor deviations do not tend to be smoothed out by the system but are instead amplified. That is why even though we may have the equation of motion of a chaotic system (such as the double pendulum), we cannot predict the eventual state of the system, because in a real-world system, there is no such thing as a completely accurate measurement, so any initial values we insert into the equation of motion will differ—even if only by a tiny amount—from the true values, and even tiny differences in initial values will lead to enormous differences as the system evolves.

The third condition tells us that chaos is not the same as total disorder. Random noise, such as radio static or the turbulence of a rushing river, is not a chaotic system. Static is totally random, and chaos is anything but. Chaos appears very irregular, but not everything that looks "chaotic" is. The third condition adds something else as well. As it swings, the path of the double pendulum will describe a dense scribble on a piece of paper, eventually coming arbitrarily close to every point within its range. Yet its path is governed by the simple principle of the pendulum's design; it is anything but random! Thus the third condition also means that a chaotic system will essentially fill the available space, in the sense that that there is no region in the range of the system, no matter how tiny, that the system will not eventually penetrate. In a sense, chaos fulfills the principle that nature abhors a vacuum.

In this technical sense of the term, chaos is not a state of utter confusion. In fact, that is the point: a system exhibiting mathematical chaos *appears* chaotic, but it is subject to a simple set of rules. There are structures that are even more complicated than chaos, which I will only mention: Brownian motion, turbulence, vortical flow. These highly complicated structures do not count as chaotic. Perhaps the most interesting thing about chaos is that it is theoretically simple.

Mathematicians and physicists often feel an aversion toward systems that are overly complex, especially since even simple systems are often unsolvable. Nature has no such aversion. It is not trying to solve anything. In the natural world, things just arise according to the laws of physics, chemistry, and biological evolution, and nature decides what will eventually prevail and what will perish, without asking whether a particular structure is too complicated.

The construction of the human brain is largely determined by the information encoded in our DNA, and although our brain contains many more variables than the five to ten required by the first condition for chaos, it nevertheless has many orders of magnitude fewer variables than the quantity of information required to describe it. Nature interprets condition (1) on a much grander scale than mathematicians and physicists do, and the thousands of genes that encode the rules to construct a human brain are, by nature's standards, a "small" number of variables. We can imagine, then, how natural selection managed to produce such an immensely complex structure defined by such a relatively small number of variables, even if mathematicians or physicists would hate to work with so many.

It appears that for the human brain to emerge with all its higher cognitive functions, it was necessary that chaos be able to operate. That does not mean the laws of chaos had to be encoded in our DNA, any more than the gravitational constant needs to be encoded in an animal's brain in order for it to keep its balance. But the laws of physics are part of the natural environment, and organisms evolve under the possibilities and constraints of those laws.

Human beings may be the only animals whose brains can put the laws of chaos to use in cognitive processes. Electroencephalographs reveal that humans are actively thinking all the time, even in sleep. No other animal is known to do this. Even our relatives the great apes exhibit periods of no activity even while awake, and then it takes an external stimulus to activate their higher brain functions. Constant activity enables a degree of long-term chaotic behavior in the human brain, and this appears to be a defining component of human thought.

The Science of Chaos

One of the principal motivations for the development of mathematics has been to find better ways of calculating, and over the past ten thousand years, this search has resulted in ever more sophisticated calculating tools and algorithms. But even the cleverest mathematics can be defeated by human irrationality. Newton, for instance, wrote of the stock market in 1720, after he lost £20,000 (a huge fortune back then), "I can calculate the motion of heavenly bodies, but not the madness of people."[6]

Since ancient times, mathematicians have also debated the nature of the objects that mathematics studies, such as numbers and geometric figures. As early as the fifth century BCE, the Pythago-reans discovered irrational numbers, and the ancient Greek geom-eters asked whether an arbitrary angle could be trisected using only a straightedge and compass, a problem that remained unsolved un-til the nineteenth century and the development of more advanced mathematical theorems that were used to show that such a construc-tion is impossible.

Even though it took two millennia to solve the angle-trisection problem, mathematicians remained firmly convinced that every mathematical problem that could be posed would eventually be solved and that every calculational problem that could be calculated in theory could ultimately be calculated in practice. This may be why Poincaré's results on chaos had so little impact in his time. But Gödel's theorem swept away the idea that mathematics could solve every problem mathematicians could state. Once it was clear that mathematics could not calculate everything and could not prove or disprove every assertion that it could make, mathematicians became interested in exploring the subject's boundaries. What things are im-possible to calculate or predict? And does mathematics have anything interesting to say about them other than that they are incalculable or unpredictable?

So far in this chapter we have mainly seen the negative aspects of chaos: the things it declares incalculable and unpredictable. Yet although we cannot predict the future state of a chaotic system, it is sometimes possible to calculate the probability that it will be in one set of states rather than another. Such a calculation provides a kind of theoretical solution, but it turns out that the nature of chaotic systems makes it impossible to predict with any degree of accuracy the occurrence of an extreme event.

Here it is important to keep in mind that chaos theory is not con-cerned with disorder, randomness, or confusion but with a precisely defined type of *apparent* disorder. This is just the way science works; whenever possible, it seeks to address simple questions—those that can be answered by experimentation—leaving the really big and difficult problems to other modes of acquiring knowledge. Science gives wide berth to questions like "What is the meaning of life?" "Why is there something rather than nothing?" or "What is the ultimate harmony of the world?" and instead asks much more mundane questions: "How fast does a ball roll down an inclined plane?" "What route does blood take within the body?" "How does animal reproduction work?" and "Why does an apple turn brown if you crush its flesh with your finger but an orange doesn't?" (By the way, it is this last, seemingly random, question that led the Hungarian scientist Albert Szent-Györgyi to discover vitamin C.)

The way chaos theory is formulated makes it simple enough for scientific inquiry, while it describes a large enough group of phenomena for its results to be widely applicable. In answering such narrow, seemingly simplistic questions, scientists are able to reach very general conclusions. For instance, the law of conservation of matter and energy could easily appear in the writings of mystics or philosophers or be rendered as a work of art. Those modes of discovering the world have actually arrived at conservation laws, though not with the fastidious precision of science. It is a characteristic of science that we not only know what we know, but also how we arrived at that knowledge.

Chaos theory has led to the discovery of scale-invariance, a principle as general and elegant as the law of conservation of matter and energy.

8

Scale-Invariance

How many times have I wished that my glasses had a phone number!

Figure 14 shows the pound-to-dollar exchange rate over various time intervals during the fiscal year 2012–2013. The first thing to notice about these graphs is that I neglected to label dates and times on the horizontal axis, and I also failed to supply the vertical scale. Can you tell which graph represents a five-minute history and which represent one-hour, one-day, and one-week histories? I will not spoil the fun by revealing the answers here; the solution can be found at the end of the book.[1] And take heart, gentle reader, if you failed to figure out which graph is which. Not even the most renowned stock market gurus can do it.

Self-Similarity
The fact that financial market graphs look the same at all time scales caught the attention of Benoit Mandelbrot, whom we encountered in chapter 6. He wanted to know whether the experts were missing something or whether there was indeed no way to tell the difference.

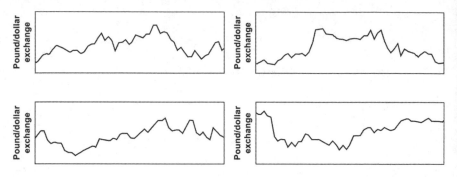

*Figure 14. Graphs of pound-to-dollar exchange rates. Which curve
is scaled to five minutes? Which one to an hour? A day? A week?
(Drawings by József Bencze.)*

If the four graphs in figure 14 showed the exchange rate between
the British pound and British pence or between the U.S. dollar and
U.S. penny, and they were therefore horizontal straight lines because
the exchange rate never varied, nobody would be surprised that you
could not tell the time scale. But the exchange rates in these graphs
are subject to intense fluctuation, and it is reasonable to suppose that
these fluctuations should have a kind of temporal rhythm such that
the changes over a minute and those over a week should look very
different. Yet they are eerily similar.

To model such a graph, Mandelbrot wanted to find a mathematical object whose scale was invariant not only in practice—to the
naked eye, so to speak—but also in theory. *One* such object obviously exists: the straight line. But are there any further, nontrivial (as
a mathematician would say), examples of such objects? If they don't
exist, then there is something yet to be discovered in stock market
curves that will one day allow us to determine the time scale of a
market graph. Such knowledge could lead to valuable new discoveries about the nature of financial markets.

If we are not looking for rigorous mathematical self-similarity but simply for objects that look the same on a variety of scales, we can see that nature provides several examples. The fern, for instance, has large leaves, each of which comprises numerous apparently iden tical smaller leaves, each of whose leaves comprises numerous apparently identical smaller leaves, and so on (figure 15). At a certain point this self-similarity breaks down: individual fern cells look like normal plant cells, not like fern leaves.

Examples also appear in art. Figure 16 shows a mosaic from the seventh-century Basilica of Santa Maria in Cosmedin, in Rome. Based on that idea of triangles within triangles, the Polish mathematician Wacław Sierpiński discovered a truly self-similar mathematical object that can be obtained by continually removing triangles through infinitely many iterations. Figure 17 shows the fourth iteration of that process.

Figure 15. A self-similar fern.

Figure 16. Seventh-century mosaic in the Basilica of Santa Maria in Cosmedin, Rome (photograph by Francesco De Comite; reprinted under license https://creativecommons.org/licenses/by/2.0/legalcode).

Figure 17. Fourth iteration of the Sierpiński triangle (drawing by József Bencze).

Other truly self-similar mathematical constructs were discovered as early as the late nineteenth century, but before Mandelbrot, they were seen mostly as curiosities. Mandelbrot called such objects "fractals," and we shall soon see why.

Fractals

In the late 1970s, Mandelbrot was working at the Thomas J. Watson Research Center of IBM and therefore had access to high-powered (in their day) computer graphics. In 1980, he wrote a computer program to display the object shown in figure 18, which has become known as the Mandelbrot set. This set, or more precisely its boundary, is defined using a relatively simple formula, and the curve that forms this boundary is scale-invariant. No matter where we zoom in, it looks like the original figure. There is no way to tell

Figure 18. The Mandelbrot set (upper left), with views obtained by successively (clockwise) zooming into the center of the picture, each time with a successive magnification factor of several billion.

at what magnification we are viewing it. If you look online, you can find some spectacular animations zooming deep into the Mandelbrot set, with the original shape appearing again and again, demonstrating its self-similarity.[2]

Needless to say, the boundary of the Mandelbrot set is no ordinary curve, such as a circular arc or even some fantastically winding twisted curve. In fact it is not a one-dimensional curve at all. Yet it isn't two-dimensional either, because it does not cover any entire segment of the two-dimensional plane. It is spread out like a very wispy cloud. If we want to assign it a dimension, it will have to be some number between one and two. This fractional dimension is what led Mandelbrot to call sets of this kind fractals.[3]

Many images of these remarkable objects, as well as programs to create them, can be found online, and I encourage the reader to explore them. Fractal generators require only a few parameters, yet they yield an extraordinary richness of form. We saw one representation of a fractal in figure 8, and two more appear in figure 19, created in the simplest way possible, using only the fractal generator of the

Figure 19. Fractals generated in Photoshop (created by Vera Mérő).

general-purpose image processor Photoshop. With the aid of fractal generators, we may also add color to make the image even more spectacular and reveal the underlying regularities and symmetries.

Scale-Invariance as a Law of Nature

Mandelbrot found that graphs of financial market activity have many of the properties of fractal curves. This settled the question whether one could determine the scale of a financial market graph. If they are fractals and thus self-similar at all scales, it means the financial experts have not missed some trick that would allow them to determine the scale; to the extent that the graph is self-similar, there is simply no way of telling, even theoretically. It seems that by their nature, financial markets are scale-invariant.

Just as the initial conditions of the double pendulum determine its path, the parameters of a fractal determine how it will develop as it is generated. We saw with the pendulum that small changes in the parameters result in wildly different paths. Does the same hold for fractals? How sensitive is their development to initial conditions? As we shall see, the answer is "extremely sensitive."

Mandelbrot developed the fractal concept to model financial market behavior, but he soon began to wonder whether fractals are the rule in nature rather than the exception. Coastlines, for example, form random zigzags that look like last week's Dow Jones average, sometimes with islands extending out like wispy clouds. From a distance, the jagged contours seem well defined, but the closer we look, the more we see their elaborate indentations, and it finally becomes almost impossible to determine whether a particular point—a pebble or a grain of sand—belongs to the sea or the coast. In fact, coastlines are just as fractal as the boundary of the Mandelbrot set.

Mandelbrot's early thinking about fractals is presented in his 1967 paper "How Long Is the Coast of Britain? Statistical Self-Similarity and Fractional Dimension." Here he describes the "coastline paradox," the fact that the shorter the ruler used to measure a coastline, the longer the measurement becomes, since a shorter ruler lets us measure more of the zigs and zags. To be sure, we achieve a similar result for a simple circular arc, but there the longer measurements with shorter rulers approach a fixed limiting value, which we call the length of the arc. The same holds for all other garden-variety curves but not for fractals, whose lengths diverge to infinity. To the extent that a coastline is fractal-like, there are essentially infinitely many segments, large and small, to measure. Mandelbrot showed that there is no precise way to define the coastline of Great Britain or its length. It has no length, just as the Cauchy distribution, as exemplified by our friend Phoebe, has no standard deviation. So fractals, like the Cauchy distribution, have brought us to Wildovia.

Mandelbrot was so inspired by this phenomenon that he began to collect examples of fractal-like phenomena in nature. He found that once we know what we are looking for, we bump into them almost everywhere. The leaves of a fern, as we saw, are fractal-like, as is the elaborate tunnel system of a mole. Mountaintops, snowflakes, clouds, and the boundaries of Norwegian fjords are also fractal-like. Even the human brain can be considered a complicated fractal. These observations led Mandelbrot to publish a book in 1983 called *The Fractal Geometry of Nature.*

Psychologists also took an interest in fractals. They investigated what kinds of images (landscapes as well as abstract paintings) people find beautiful, and one consistent result was that people are attracted to fractal-like images.[4] Perhaps it is because we are so surrounded by fractals that these images seem more familiar than those of a more conventional geometry. What is surprising is that it took so long for psychologists to discover this fact, since fractals are just about everywhere.

Fractal-like images have been present in art for a long time, such as the seventh-century Sierpiński triangle mentioned above. We might also mention the flamboyant arches and tracery of Gothic architecture, which exhibit some degree of self-similarity, as do many modern paintings. And in the years since fractal-generating programs became widely available, a whole new genre of art has emerged that consciously makes use of fractals. Mandelbulb, shown in figure 20, was created by Daniel White and Paul Nylander and is based on a three-dimensional variant of the Mandelbrot set.

*Figure 20. The Mandelbulb
(created by Daniel White and Paul Nylander).*

Today's computer graphics artists use fractals intensively. Every hillside and every cloud in your favorite video game is constructed by a fractal-generating algorithm that produces lifelike images. There is even self-similarity in literature: in a heroic crown of sonnets, the final binding sonnet, the concise summation of what came before, consists of the initial lines of the fourteen previous sonnets. In music we have the fugue, which exhibits self-similarity in the repeated appearance of a single theme, and it also shows scale-invariance in the use of augmentation and diminution, whereby the theme appears in larger (augmented) or smaller (diminished) note values; stretto, with one iteration of the theme chiming in before the previous one has finished; and inversion, in which the theme appears upside down.

For engineers, scale-invariance could have enormous advantages, since a single design could serve to produce a piece of machinery that performs a certain function in every possible size. But we would immediately run into difficulties, such as the fact that as identical three-dimensional objects grow in size, they do not maintain the same ratio of volume to surface area. This can lead to problems of structural or thermodynamic stability. Nature, on the other hand, doesn't *design* anything. It just tinkers with things, and whatever survives, survives.

If we were to discover a law of nature to the effect that everything strives to be as scale-invariant as possible, that would represent a huge step in our understanding of how structures of incredible complexity arise in the natural world. It would suggest that things become scale-invariant not because of any design principle specific to their individual history but through a general law. If this hitherto unknown general operating principle were found, we could credit Mandelbrot with its discovery. But if there is such a principle, we know very little of the mechanisms that make it work, and we are even less able to determine its domain of validity.

Scale-Invariant Chaos

Chaos and scale-invariance go hand in hand. The only exception is the obvious and trivial line segment. All other scale-invariant objects possess all three characteristics of chaos presented in the previous chapter.

1. The system should be defined by a small number of variables. The Mandelbrot set, for instance, is defined by a very simple equation with a single complex variable, and even the Mandelbulb shown in Figure 20 is defined by only three variables. If we allow randomness so as to increase richness in form, that adds only one more variable. More complicated fractals are defined by more equations, but the number is usually in the range of five to ten. But even fractals created with a much larger number of variables can exhibit chaos, which is what we saw with the human brain: it is created from thousands of genes, and it exhibits chaos.

2. The system should be very sensitive to small differences in the initial state. In the case of a fractal, the initial state is represented by the equations that define it. And sure enough, a minute change to the parameters of those equations radically changes the fractal's appearance.

3. A chaotic system should find itself at some time in its evolution arbitrarily close to every state that it is theoretically capable of reaching. In the region of the plane or three-dimensional space (or multidimensional space) in which it is defined, a fractal is dense in the sense that a cloud is dense: it is not solid, but it comes close

to every point in its region of definition. Whatever points in that region do not belong to the fractal are arbitrarily close to points that do belong to the fractal.

The unpredictability of chaos also applies to fractals. If we take a random point in the plane and ask whether it belongs to a given fractal, there is no universal way to decide. This may seem counterintuitive, because the fractal is defined by a few equations, so we should theoretically be able to determine whether any one point belongs to the set of solutions. But Gödel tells us that we should not be surprised if we cannot do so. With the double pendulum, we can follow its path from the initial conditions, and if it passes through our randomly chosen point, then we can conclude that the point was in the path. But if the pendulum has not crossed our point, we cannot predict whether it ever will.

The same holds for fractals: the only way to decide whether a particular point belongs to a given fractal is to let our computer keep generating the solutions to the equations. If the computer draws the exact point we have chosen, then we can be sure that it belongs to the fractal. But until this happens, we have no idea whether it will ever happen. A consequence is that if our point does not, in fact, belong to the fractal, we will never find that out, no matter how long we let the computer run.

Although all fractals are chaotic, not every chaotic phenomenon has a fractal-like structure. For example, the trajectory of the double pendulum is chaotic, but it is not a fractal. It is true, however, that fractals are the commonest manifestations of chaos in nature. That is, chaos usually appears in nature in a scale-invariant form. That would not be surprising, of course, if it turns out that Mandelbrot in fact pointed out a hitherto unknown principle that is valid under very general conditions. Scale-invariance may be one method by which nature can economically create objects with an extremely rich structure. It could also be that scale-invariance is how nature realizes its abhorrence of vacuums. Unless we are talking about the trivial case of a line segment, scale-invariance automatically creates chaos, and as we saw in the previous chapter, chaos abhors a vacuum in the sense that it densely fills its space of definition. Based on what we know today, this is all speculation, but what we know for sure is that scale-invariance and its counterpart, chaos, can be found throughout nature.

Scale-Free Networks

Although Mandelbrot was interested in self-similarity as primarily a geometric phenomenon, scale-invariance has proved to be a much more general concept. One of its most fruitful applications was the discovery of scale-free networks, made world famous by the Hungarian-American physicist Albert-László Barabási in his best-seller *Linked*.

To mathematicians and physicists, a *network* is a structure that consists of a collection of nodes *(vertices),* some—but not necessarily all—of which are connected by *edges*. Networks can represent all kinds of relationships. For example, to show personal relationships among a group of people, each person can be represented by a node, and then an edge between a pair of nodes might show that those two individuals are acquainted. The neurons of the brain also form a network; some are connected, others are not. Web pages form a network, the so-called web graph. An edge joins two pages if one page has a link to the other. One could also draw a network of scientific research publications whereby a link represents one publication citing the other. Airline routes form a network; the nodes are cities, and there is an edge between two cities if they are connected by a direct flight. The big discovery made by Barabási and his collaborators was that most networks found in nature, as well as social networks, are more or less scale-invariant, just as most naturally occurring chaotic systems are scale-invariant.

Some networks have a certain asymmetry. For example, in the network of scientific papers, if paper B cites paper A, then it is extremely likely that paper A does not cite paper B (since B will generally have been published after A). The edge linking nodes A and B thus has a direction. We call such networks *directed,* and we can speak of an edge at a node as either incoming or outgoing. Similarly, in the airline network, there may be a direct flight from Altoona to Pottsville but not from Pottsville to Altoona, so there is an edge running from the Altoona node to the Pottsville node but none in the other direction.

Scale-invariance in networks is similar to the scale-invariance of geometric figures: every part of the network looks basically like every other part, subnetworks look like the whole network, and subsubnetworks look like the subnetworks in which they reside, so it is impossible to tell at what scale we are viewing the network. If we view a network at a different scale, taking towns or counties as vertices instead of individual people, its appearance doesn't change much.

Scale-free networks have interesting properties that are not found in most other networks. For instance, scale-free networks are very dense in the following sense: every pair of nodes can be reached by a relatively short path. For example, it is estimated that any two people on earth are connected through a path of at most six acquaintances. The Internet also forms a very large network, in this case a directed network, with about one page for every two people on Earth. Here, too, almost every page can be reached from any other page by at most twenty clicks. (Of course, just as there may be small groups of people with no outside acquaintances, there are pages with no incoming links. Although such "islands" cannot be reached from the outside, what we have said about connectedness applies to the great majority of Web pages.)

Another property of scale-free networks is that compared with "normal" networks, they have a relatively large number of nodes with many more incoming or outgoing connections than the average, while most nodes have relatively few connections. It is through these well-connected "hubs" that most information is transmitted in a scale-free network. If we want to spread a piece of information in a scale-free network, we first have to locate the hubs. In the social sciences, these hubs are often called opinion leaders. Among certain primates, they are often older females, whose job is to provide grooming services to the entire group, spreading information as they move about. In human societies, the postman or hairdresser may fulfill a similar role.[5]

An especially interesting case of a scale-free network is a possible model of how we look for something that we have lost. We usually rummage about in one particular spot, taking only small steps while doing so. But after a while, we suddenly move away and begin searching in an entirely different place, where we again forage about with small steps. If we draw a network whose nodes are the spots where we search, with the edges representing the paths we took between those spots, we get a scale-free network called a Lévy flight, named after its discoverer, the French mathematician Paul Lévy.[6]

Let's suppose we are looking for our glasses or our cell phone (although at least we can call the cell phone if there is another phone around; how many times have I wished my glasses had a phone number!). If we can't call our phone, we will look for it in the manner described by Lévy. Scale-free Lévy flights are also used by bees and albatrosses, deer and swallows, in searching for food and perhaps also for nesting material.

Paul Lévy described this search algorithm back in the 1930s, and he proved that under certain circumstances, it is the optimal search method. The reason is that this strategy minimizes the possibility of revisiting already searched areas while maximizing the number of visited locations. Thus Lévy proved that scale-invariance can have theoretical and even practical advantages—only he didn't call it scale-invariance or self-similarity, because those concepts had not yet been discovered. He simply recognized the existence of a very special parameter, one that also plays a fundamental role in the science of Wildovia.

John von Neumann was impressed by Lévy's thought process. Mandelbrot wrote, "As a later teacher of mine, John von Neumann, told me, 'I think I understand how every other mathematician operates, but Lévy is like a visitor from a strange planet. He seems to have his own private methods of arriving at the truth, which leave me ill at ease.'" As for Lévy, Mandelbrot added, "When later I told Lévy how I developed his ideas and applied them to economics, he was flabbergasted, and, perhaps, annoyed. In his view, 'real' mathematicians simply did not do such prosaic things as study income or cotton prices."[7]

The Components for the Emergence of Wildovia

Lévy set out on the trail of scale-free networks by solving a purely mathematical problem, but it was another half century before that concept would enter the world of scientific research. One point of interest about scale-free networks is that we have a good idea how they can arise spontaneously. Albert-László Barabási and Réka Albert devised a very simple and elegant mathematical model to illustrate this, and they tested it on a variety of real-world networks, including the network of Hollywood actors linked by having appeared in the same film, certain parts of the Worldwide Web, and the United States electrical grid. In every case, the networks followed their model with fairly good accuracy.[8]

Let us imagine a network being constructed one step at a time, with newcomers connecting preferentially with members who were the earliest to join; the earlier a person joined, the more likely it is that a newcomer will link to that person. This means that early members gain an ever-increasing connection advantage over late arrivals. Barabási and Albert proved that under certain conditions, this simple principle suffices to produce a scale-free network. The same result obtains whenever certain elements of the network are preferable to others, for whatever reason. (The situation becomes more interesting, and more complicated, if we consider the possibility of strong and weak ties, but I will not pursue that here.) This effect, whereby preferential attachment leads to a scale-free network, is called the Matthew effect, after the biblical verse "For unto every one that hath shall be given, and he shall have abundance: but from him that hath not shall be taken away even that which he hath" (Matthew 25:29).[9] In modern parlance, the rich get richer, and the poor get poorer.

In addition to the Matthew effect, three other phenomena have been found that may contribute to the emergence of scale-free networks.[10] The first is an increase in complexity, which usually encourages a network to assume a modular structure, which in turn can lead to scale-invariance. The second is the process of accumulation—for example, the accumulation of knowledge or capital. The third is intense competition, as exemplified, for instance, in biological evolution, which has led to the emergence of some organisms with outlandish features. While there is as yet no formal proof describing how any of these components would lead to the emergence of scale-invariance, there are several logical arguments that suggest such influence, as well as a certain intuition that increasing complexity, accumulation, and intense competition contribute to the creation of scale-free networks in nature as well as in human society.

Scale-invariance means chaos: minute changes in initial conditions lead to enormous differences in how a scale-invariant network will develop. That is why we can find an extremely rich variety of natural and social networks, even though the underlying principles are relatively simple.

The Strength of Weak Ties

Back in the 1960s, the American sociologist Mark Granovetter was studying how people conduct job searches. After analyzing hundreds of interviews and questionnaires, he was surprised to find that most people did not land a job through a newspaper advertisement or close acquaintance. In almost 80 percent of the cases, the key to success was an acquaintance whom the job seeker knew only slightly. When Granovetter published his now famous paper in 1973, he titled it "The Strength of Weak Ties."[11] That paper has become a large hub in the network of sociological publications, with around thirty thousand citations.

Barabási found that this phenomenon was not limited to job hunting. Rather, it emphasizes one of the most intriguing characteristics of scale-free networks: almost all of the many connections in a hub are weak. Paradoxically, it is precisely these weak ties that prevent the network from falling apart.

In scale-free networks, strong ties form islands. The members of such an island spend most of their time with their fellow nodes and may be largely shut off from the rest of the network. These islands are connected to the rest of the network by weak ties. In a hub that contains many islands held together by many weak ties, it is those weak ties that hold everything together, thus preventing the network from falling apart. That is why Granovetter could observe that for the most part, it is not close friends who help a person find a job. Close friends know mostly the same people as the job seeker, and they tend to suggest people whom the job seeker has already contacted. If we had only strong ties, we would find ourselves marooned in a very closed world.

The Hungarian biochemist Péter Csermely studied stress proteins for many years. These are proteins that form an organism's most ancient defense system. Whenever a protein folds the wrong way, the stress proteins stretch it out, giving it another chance to fold properly. Since proteins can assume a variety of three-dimensional structures, they occasionally fold in a way that prevents them from performing their proper function. Csermely writes, "Without stress proteins, the cell would overflow with a muck of misshapen proteins holding on to each other like there was no tomorrow." The big question Csermely answered was how stress proteins managed to be on hand when they were needed. Csermely continues: "In the first five years I attacked [the stress proteins] with everything a biochemist might try. I isolated them, chopped them up, cooked them, and soaked them in acid, alkali, and radioactive goop. It took me five years to find out that stress proteins weren't like their fellows. . . . Not only do stress proteins twist, they also stick, but just a *little bit*, yet in the same way to everything."[12]

Surprisingly, the key to understanding how stress proteins work wasn't to be found in biochemistry but in network theory. Stress proteins act like hubs in a scale-free network. These are the proteins with many weak ties. Other proteins, busy with their particular physiological functions, are strongly tied to a few others, with which they perform some physiological function, and they have no time or energy for sustaining weak relationships. Stress proteins are like older females in primate societies who groom the whole tribe and thus keep it a cohesive unit.

Scale-free networks are usually stabilized by weak ties, and this sort of stability is characteristic only of scale-free networks. It is what allows them to remain fairly constant over long periods of time, even when they grow so huge that no vertex is connected directly to more than a negligible portion of the entire network. It is through weak ties that a metropolis with a population in the millions can function as a coherent unit. Weak ties allow the hundred billion neurons of the human brain to produce consistent thoughts. And it may be that weak ties underlie the pursuit of scale-invariance as a fundamental principle of nature.

9

The Levels of Wildness

There is something more chaotic than chaos.

Let us imagine that we live in the Land of a Million Lakes. We have lakes of many sizes. The largest is seventy-five miles wide. The second-largest is forty-five miles wide, the next-largest after that, thirty miles. Even the thousandth-largest is several hundred yards wide. There are, of course, hundreds of thousands of ponds only a few yards across, but those don't really interest us.[1]

Our lakes have been thoroughly mapped. We know the dimensions of every one of them. Across the border, however, the Land of a Hundred Million Lakes is uncharted territory. It is a lot like our country, only a hundred times larger. We are explorers setting out to cross an unknown lake in this undiscovered country. We cannot see the other side in the fog, yet we set out bravely in our little rowboat. We believe that we have enough strength and enough food. It is very unlikely that we are so unlucky as to have chosen a lake that is one hundred miles wide, although such a huge country could have even larger lakes than that. Knowing the lakes back home, we figure that even if this lake is relatively big (and it could be, since we cannot see the other side even though the fog is lifting), still it is probably not much more than five miles wide, and crossing it would be a piece of cake.

Time passes, and we have rowed twenty miles and still cannot see the other side. Feeling somewhat discouraged, we ship our oars for a moment to reassess the situation. Just how far do we have left? If we are pessimistic enough to believe we might be dealing with the phenomenon of eternal youth, an extreme phenomenon of Mildovia, then we should expect to have the same distance remaining that we expected at the start of our journey, namely, another five miles.

If we are, in fact, in Mildovia, then this is a very pessimistic point of view, since eternal youth is not something that occurs every day in Mildovia, and only a few things are forever young there. More typical of Mildovia is that things age, so in our case, this would mean that the more we have rowed, the less we should expect to have left. That seems much more likely. But Wildovia has much weirder things than eternal youth. If it should turn out that we are in Wildovia, then the more we have rowed, the *more* we should expect to have left. That is not a very happy state of affairs for us in our little boat!

So with a certain trepidation, we start counting. Let's suppose the distribution of the widths of lakes in this huge country is the same as at home, only the population of lakes is a hundred times greater. Perhaps that distribution was already Wildovian, but we didn't realize it because we knew all the lakes by name.

Gloom sets in as we realize that at home as well, the more we rowed, the more we were likely to have left. For example, if we have rowed thirty miles in one of our homeland lakes and still cannot see the other side, we may conclude that we are on one of two lakes— the only two that are wider than thirty miles. We must have either fifteen or forty-five miles left. So at home, after thirty miles, we have on average thirty miles more, and not five as we originally thought. We also realize that in our homeland, this is true not only of the two biggest lakes, but also at any point of the journey. Considering all of our familiar lakes, we sadly realize a general rule: if we have rowed x miles, we may expect that another x miles remain.

The distribution of lakes in our homeland is Wildovian in the sense that the more we have rowed, the more we probably have left. Thus it is not only on our adventure far from home that we face a much worse outlook than the extreme Mildovian situation of eternal youth, but—we now realize—even in our homeland.

Since the geographical conditions in this huge country are basically the same as in our home country, we have good reason to suspect that the distribution of lakes is the same as well, but with many more lakes. The more we have rowed, the more we expect to have before us. But how much more? How much do we expect here, in a huge land, if we cannot see the shore even after seventy-five miles?

At home, if we have gone that far, we know that we must have arrived at the shore, and if we don't see it, we had better start worrying about our mental status. But in the Land of a Hundred Million Lakes, nothing guarantees that the biggest lake is a mere 75 miles wide. In such a vast country, there could be lakes hundreds, even thousands, of miles wide. Out in this wilderness, we are forced to concede that we expect 75 more miles ahead of us. And if after rowing another 75 miles we still do not see the shore, we shall have to expect another 150 miles of rowing ahead of us.

The Mandelbrot Factor

But were we justified in our estimate of a simple proportionality? If the distribution of lake widths is well modeled using the mathematics of scale-free networks, corresponding to the distribution of outgoing connections from each vertex, then our estimate is justified. Mathematics says that if a scale-free network proves to be a good model for the distribution of lake widths, then that is the way to compute. It needn't concern us right now whether this model actually describes the distribution of our lakes. Since we invented the Land of One Hundred Million Lakes, let's just take it for granted that it does.

Consider an acquaintanceship network. Suppose each person in such a network has on average 150 acquaintances. How many acquaintances should we expect of someone who is known to have more than that? Our experience with reaching the end of a lake corresponds to reaching the end of a person's list of acquaintances. We infer, then, that in addition to the 150 we know about, we can expect another 150, making it 300 in all. And if we know that someone has more than 500 acquaintances, we should guess that he or she has at least an additional 500.

Even using the model of a scale-free network, there is a parameter we haven't considered: in the case of interpersonal relationships, the proportionality coefficient might not be equal to 1, as it was with our imaginary lakes. It might be more or less, depending on how we constructed the distribution.

To take another example, research into the network of human heterosexual sex partners has shown this coefficient to be closer to 2, possibly even a little higher, although the acquisition of accurate data is made difficult by the fact that men report an average of seven sexual partners, while women report only four. The number should be the same for both sexes; it is likely that men overstate it and women understate it. We might also note that prostitutes and "sex addicts" were included in the sample, and their inclusion may have skewed the results. But only somewhat, for research has shown that the large difference in reported numbers is a genuine cognitive bias in both sexes.[2]

In any case, a coefficient of about 2 has been found to be valid for both men and women. Hence, if we know that a person has had at least four sexual partners, we can expect that he or she has had an additional eight. Note that this is just an expected value; a total of four might actually be correct, but so might ten or more; eight is simply the average number of additional partners. And if we know that someone has had at least ten partners, then that person has had (on average) about twenty more. It is interesting that if we leave out prostitutes and those suffering (or not) from nymphomania or satyriasis, the proportionality coefficient of the network of sexual relationships drops to about 1.

Proportionality coefficients are a characteristic of scale-free networks, and they can be any positive number. Let us call the proportionality coefficient of a scale-free network the *Mandelbrot factor* of the network. Thus every scale-free network has a Mandelbrot factor, and that number determines the network's main characteristics. Mathematicians usually use an exponent rather than a proportionality constant to describe a network, because exponents are easier to calculate with.[3] But to avoid getting into higher mathematics, we will stick with our proportionality coefficient.

All of this applies not only to scale-free networks, but also to all kinds of scale-invariance, such as the lands of lakes: the larger the Mandelbrot factor in a land of lakes, the more distance we must expect ahead of us before we get to the other side.

Scale-Free Incomes

Let's assume for a moment that incomes (at least the really high ones) are scale-invariant. Suppose we know that a certain Ms. Fortunato has an annual income of at least $1 million, but we don't know exactly how much. If the Mandelbrot factor for high incomes is 2, then we can expect our wealthy acquaintance to have an additional $2 million of income, or $3 million altogether. (But keep in mind that these numbers are expected values. Ms. Fortunato might earn only $1 million after all, or she might rake in $10 million or more.) And if we know that Mr. Rich has an income of at least $10 million, we should expect his income to be around $30 million, with, again, a wide range of possibilities.

At this point I have to apologize to Vilfredo Pareto, whose formula I described in chapter 5 as resembling the work of a charlatan. In fact, the Mandelbrot factor is uniquely defined by the formula that Pareto used to try to describe the distribution of incomes (though the actual conversion is rather complicated). I noted that Pareto's formula, unlike the lognormal distribution, is not connected in any meaningful way to other mathematical achievements and is a totally artificial construct. If scale-free networks had been known in Pareto's time, that judgment would have been utterly unfair. Nevertheless, Pareto's formula works painfully badly on below-average incomes, so perhaps we should maintain our harsh judgment even if Pareto unwittingly anticipated the science of scale-free networks.

Extremely high incomes, however, are better approximated by Pareto's formula than by the lognormal distribution, which we discussed in considerable detail in chapter 5. Low and medium incomes are well modeled lognormally, and even relatively high ones are lognormal to a good approximation, but extraordinarily high incomes are not. This phenomenon hints that while the majority of incomes are Mildovian, high incomes work according to the laws of Wildovia.

The American economist Edward P. Lazear gave a surprising explanation for this phenomenon.[4] Extremely high incomes, he argued, are determined by entirely different factors from those that govern lower ones. The income of a CEO is not $10 million a year because he is so highly profitable to the firm. It is set high in order to promote competition among upper-level executives—because one of them, perhaps, will someday become the new CEO—thereby motivating them to maximum performance. Thus the CEO does not receive an astronomical paycheck because he himself has earned it, or even because he personally motivates the other employees, but as a reward for having won the competition to be top dog. If the CEO's high salary actually motivates those below him, then it is worth it to the shareholders to pay him so much, even if he does little to promote the interests of the firm, though due to the virtues by which the CEO attained his lofty position, he will likely serve the firm's interests in any case.

Even if Lazear is correct and CEOs are given large salaries to promote competition in the levels of management just below the top, there is likely another force also pushing those salaries to astronomical levels. As we saw in the previous chapter, extreme competition alone can easily bring about Wildovian conditions. The competition generated by somewhat high CEO salaries can lead to even higher CEO salaries, leading to even more competition, leading to yet higher salaries. At the same time, most workers' salaries are set by the laws of Mildovia and are therefore distributed lognormally, while top salaries, produced by the laws of Wildovia, are governed by scale-invariance. Perhaps that is what the Swiss electorate had in mind in November 2013 when it voted in a referendum to reject a law limiting top executive salaries.[5] The voters must have seen the proposed law as an attack on the laws of Wildovia and decided that passing it was as foolish as trying to pass a law to reduce the boiling point of water.

Between Gauss and Cauchy

At the beginning of our book, when we first met the worlds of Mildovia and Wildovia, we used the Gaussian distribution to characterize Mildovia and the Cauchy distribution for Wildovia. In figure 5, we compared graphs of the two distributions and saw that the Gaussian curve approaches the x-axis much more rapidly than the Cauchy curve, that the tail of the Cauchy curve is much fatter than the Gaussian tail, and that the Cauchy graph is thinner and more pointed in the middle.

Let me repeat figure 5 here so that you don't get your book all dog-eared by thumbing back and forth (figure 21). We noted above that the radical difference between Mildovia and Wildovia was the result of a seemingly slight mathematical difference between two distributions. It may have occurred to you then that if we could create such different worlds from two simply described mathematical curves, why not draw a third curve somewhere between the two, creating a world whose properties would lie between those of the Gaussian and Cauchy distributions? If this thought did occur to you, you

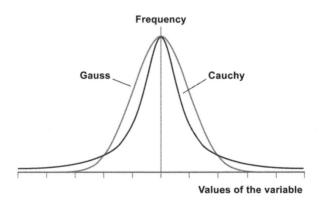

Figure 21. Figure 5 repeated: comparison of the Gaussian and Cauchy distributions (drawing by József Bencze).

have demonstrated a good feeling for how mathematicians think. In fact, the distributions that describe the connections between vertices of scale-free networks all fall somewhere between Gauss and Cauchy. If Mildovia is characterized by the Gaussian curve and Wildovia by the Cauchy curve, then scale-invariant mathematical objects lie somewhere between those two.

The smaller the Mandelbrot factor of a scale-free network, the more closely the distribution of the connections of its vertices approximates the Gaussian distribution. In other words, a smaller Mandelbrot factor corresponds to a milder network. Still, a scale-free network can never be so mild as to become predictable; it always remains chaotic. A mild scale-free network describes a relatively mild chaos. Conversely, the greater the network's Mandelbrot factor, the more closely the distribution of the connections of its vertices approximates the Cauchy distribution. This means that wilder networks have bigger hubs than milder networks.

The transition between the Gaussian and Cauchy distributions becomes especially interesting if we ask whether an intermediate distribution has a standard deviation. Recall from our earlier discussion that the Gaussian distribution has a standard deviation, while the Cauchy curve does not. It is a mathematical fact that scale-invariant distributions with a Mandelbrot factor less than 1 have a well-defined standard deviation, while those whose factor is 1 or more do not.[6] This fact shows that there really are scale-free networks throughout the range from mild to wild. Nevertheless, every scale-free network is chaotic.

In chapter 7, "The Mathematics of the Unpredictable," I defined chaos in a very narrow way and mentioned that there exist objects that are more chaotic than those that satisfy our special definition of chaos. The same applies to scale-free networks. If, for instance, the number of connections out of each vertex were determined by our markswoman Phoebe, then the network would no longer be scale-free, nor would it be chaotic in our sense of the term. It would be something much messier. As we shall see, there exist real-world networks that are not scale-free and are much more chaotic than those that are.

Milder Chaos and Wilder Chaos

The scale-invariant world is chaotic by nature, so it definitely does not belong to Mildovia. And as we have seen, scale-invariance applies not only to networks, but also to clouds, snowflakes, mole tunnels, ferns, gothic architecture, financial markets, and many other natural and social phenomena. Compared to Mildovia, the scale-invariant world is chaotic, unpredictable, and extreme even in its mildest form, namely, in situations in which the Mandelbrot factor is close to 0. At the same time, the wildest cases of scale-invariance, with a Mandelbrot factor of 2 or more, are still relatively mild forms of Wildovia. At least there is some governing principle—scale-invariance—at work.

The scale-invariant world is a kind of *mildly wild* world, where the laws of Mildovia no longer hold, yet the full wildness of Wildovia is held somewhat at bay by an organizational principle. Furthermore, this mildly wild world has milder and wilder parts. The milder parts have a Mandelbrot factor less than 1, while the wilder parts, where the factor is greater than 1, don't even have a standard deviation. If the Mandelbrot factor α is less than 1, then as we explore the connections of a given node and their connections, and the connections' connections, and so on, the percentage of connections known by us will increase and asymptotically approach 100 percent.

The smaller the value of α, the faster our knowledge of the network approaches 100 percent.

If $\alpha = 1$, then our knowledge about the node will remain constant; the percentage of unknown connections will remain roughly constant, with the number of new nodes discovered being approximately equal to the number of nodes explored.

If $\alpha > 1$, then the more connections of a node we explore, the larger becomes the percentage of unexplored nodes. The percentage of connections known by us will decrease and asymptotically approach 0 percent because the node will reveal newer and newer nodes, mostly unknown to us, faster than we can explore them. The greater the value of α, the faster our knowledge approaches 0 percent.

The Mandelbrot factor is a very elegant mathematical concept. Theoretically, it provides a precise measure of the "wildness" of a given network. Unfortunately, for any particular network, this factor is very difficult to compute, because it requires a huge amount of data, and in naturally occurring networks, the data can be inaccurate and contradictory. Nevertheless, some researchers have taken up the challenge, and several recent scientific articles have described efforts to estimate the Mandelbrot factor for a real-world scale-free network. Most of these articles give an estimate of some other parameter of the network that can be used to determine the Mandelbrot factor.

The first row of table 1 shows a few of these results. Since the actual values of the Mandelbrot factor cannot be given with a high degree of accuracy, I did not include any specific numerical estimates. Instead, I divided the networks into three groups: those whose Mandelbrot factor is well below 1 (mildly chaotic), those with a Mandelbrot factor around 1 (on the borderline of having a standard deviation), and those whose Mandelbrot factor is well above 1 (and that hence have no theoretical standard deviation). Since the estimates are rough, these groupings are to be taken with a large grain of salt. Nonetheless, this table gives a good idea of how wild the chaos is in various fields.

I have listed in the second row of the table a few phenomena that are not networks but whose distribution is approximately scale-invariant, like the lakes at the beginning of this chapter. These examples show that scale-invariance appears in many other forms in addition to the structure of networks and the enchanting geometric properties of fractals. We can see from the table just how chaotic certain phenomena in biology, social interactions, geology, technology, and economics can be. The biological food chain, for example, is not Mildovian, but it is only very mildly chaotic. The network of human sexual relationships resides at the other extreme. It is even more chaotic than most people would guess—although if we exclude prostitutes, satyrs, and nymphomaniacs, the remaining network fits comfortably into the second group, which is chaotic enough.

Perhaps surprisingly, the Mandelbrot factors of natural and human networks are frequently close to 1. Note here that the Pareto principle (or 80–20 rule), which I described in chapter 5, holds in the case of a Mandelbrot factor of 1 (or a little less). Hence our current results give even more support to the 80–20 rule and also help to define its domain of validity.

As for Mandelbrot factors greater than 1, several networks and phenomena that were once thought—for good theoretical reasons —to be scale-invariant turn out to belong to the category of the even more extreme. For instance, the world of human talent is correctly assigned not to the mildly wild world but to the truly wild world, and we use the Cauchy distribution to describe it. There is an amazing variety of ways to be talented, and the degree of talent an individual may have covers a surprisingly large range.

Table 1. Orders of magnitude of the Mandelbrot factors
of certain networks and phenomena

	Mandelbrot factor well below 1	*Mandelbrot factor around 1*	*Mandelbrot factor well above 1*
Networks	Biological food chains	Scientific collaborations	Joint occurrence of words (in every language examined)
	Sent emails	Metabolism network of the *Escherichia coli* bacterium	Citations from scientific articles
	Partnerships between companies		
	Interactions between human proteins	Network of interactions between yeast proteins	Interconnections inside a microcircuit
	Received emails	Outgoing phone calls	Human sexual relationships (including those of prostitutes, satyrs, and nymphomaniacs)
		Internet links	
		Human sexual relationships (excluding those of prostitutes, satyrs, and nymphomaniacs)	Electrical grid of the United States
		Joint appearances of Hollywood actors in a film	
		Incoming phone calls	
		Market comovements	
Phenomena	Extremely high incomes in Sweden	Shared family names (United States)	Intensities of earthquakes
	Intensities of solar flares	Extremely high incomes in the United States	Diameters of lunar craters
	Intensities of wars	Hit counts of websites	Copies of best-selling books sold in the United States
		Fatalities of terrorist attacks	

Main sources: Newman (2005), p. 8; Csermely (2009), p. 27; Taleb (2010), p. 263. The mean square error is also supplied for some pieces of data in Newman (2005).

The same sort of wildness was discovered for contact lists of email accounts; they form a network that is not scale-free, but even more chaotic. This is particularly interesting because we see in the table that the network of actual email *exchanges* is scale-free, with a not particularly large Mandelbrot factor. It is the contacts with whom we don't actually correspond but that somehow found their way into our contact list that form a much more chaotic network. (This wilder network does not appear in our table.) Other phenomena that are too wild to qualify for the table include the sizes of forest fires and the populations of bird species in America. Each has a distribution that is wilder than scale-free.[7]

Mildovian Life in Wildovia

Wildovia has relatively mild—or mildly wild—territories, which, in turn, can be mildly wild to various degrees, a range that is expressed by the Mandelbrot factor. Some of these territories are sufficiently mild that phenomena even have a standard deviation, while others are so much wilder that standard deviations are out of the question. But the mildly wild region still has a general governing principle: scale-invariance.

But in the wildly wild regions, not only are statistical fundamentals like standard deviation inapplicable, but the phenomena are not even scale-invariant. We currently know no general governing principle that could help us quantify them even approximately. Some of them are modeled rather well by the Cauchy distribution, but that is already too wild for making useful predictions.

In chapter 7 we saw in chaos theory a ray of hope for making predictions. Although the events of a chaotic system are unpredictable, the probability of a certain event occurring within a given time frame can be computed. But the probabilities of unusual events turned out to be too small for practical use, so we gave up hope of making useful predictions. Yet chaos theory is useful in helping us come to terms with the possibility of black swans popping up and preparing us to react when such swans appear. And appear they will, beyond doubt. But remaining constantly on guard against such events would make a mess of our everyday, more or less Mildovian, way of life. That is why this book is about all types of miracles rather than just black swans. We have a few millennia of experience in handling lesser miracles, so we are much better prepared for them than for black swans. The models of Mildovia and Wildovia, which provide some insight into how these realms work, help prepare us to handle miracles while we continue to live our everyday, generally miracle-free lives — mostly in Mildovia, at times along its edges, only occasionally wandering into the mildly wild parts of Wildovia.

Minor and Major Miracles

An important characteristic of black swans is that they have an enormous impact on the world. This is the fundamental difference between a black swan and most other miracles. Although a miracle is a unique and unrepeatable event, it need not have a great impact on the world. There are minor miracles and major ones as well as some occurrences that we haven't even noticed as miracles.

In chapter 3, I distinguished three types of miracles. We looked at some concrete examples of the first type, the typical miracles of Wildovia, which we called *pseudomiracles*. Some of them can be explained fairly well by scale-invariance, while others are better modeled by the Cauchy distribution. On the other hand, for the *true* and *transcendent* miracles, Wildovian conditions are unnecessary, but neither can science provide an explanation for them — due to the current state of science in the case of the true miracles and due to the very definition of transcendent miracles.

A miracle of any of these three types can be minor, major, or even world-shattering. Moses, on the exodus from Egypt, needed a miracle as enormous as the parting of the Red Sea, while for an acquaintance of mine in a moment of crisis, finding his lost car keys on a vast sandy beach was sufficient. The two are obviously not of the same order, yet we feel that both are miracles. It doesn't affect the nature of a miracle whether it is big or small or even whether we notice it.

The scale-invariant world, that mildly wild part of Wildovia, is of special importance, because there, the order of magnitude of an event improbable enough to be called a miracle (that is, a pseudo-miracle in our terminology) is characterized to a good degree by the Mandelbrot factor. For example, in the Land of a Hundred Million Lakes, a pseudomiracle of average order of magnitude can be shown to be a lake five hundred miles wide.

Whenever I give a talk on Wildovia and the logic of miracles, someone in the audience always asks the following question: If an ever-larger portion of the world today behaves according to the laws of Wildovia, miracles are obviously bound to become more and more frequent. But after a time, won't they no longer be considered miracles at all? And won't that return us to the miracle-free world of Mildovia? This is a logical argument, but it demonstrates an incomplete understanding of scale-invariance. Miracles won't become more frequent, but there will be a wider range of phenomena that are considered miracles, and what today counts as a miracle may by tomorrow be an everyday occurrence. The five-hundred-mile-wide lake that was an enormous miracle to those of us reared in the Land of a

Million Lakes might still be a miracle in the Land of a Hundred Million Lakes, but it will be a relatively unimportant one, and perhaps, as we explore the terrain and become acquainted with many more lakes, we will no longer consider it a miracle at all. But if we discover even bigger countries, lands of a billion lakes or a trillion lakes, then sooner or later we are bound to discover thousand-mile-wide lakes.

Take Facebook, which over the years has grown by many orders of magnitude and dramatically changed its role in many people's lives. The role for which it was first designed—maintaining contact with close and distant acquaintances—has remained, but for many, it has become a world of its own, the primary source of information and center of their social lives. This transformation was made possible by the fundamentally scale-invariant nature of the Facebook network.

The Mandelbrot factor of a network or other phenomenon generally does not change. It remains constant. What can change is the range of sizes of the objects we can actually encounter. But scale-invariance means, by definition, that the structure of such a network is the same on a large scale as on a small scale. So if a scale-free network grows by an order of magnitude, the frequency of miracles will not grow; rather, what changes is the threshold for considering something a miracle.

Of course, this conclusion applies only to pseudomiracles and only to those that arise in the mildly wild regions of Wildovia. In the true Wildovian wilds—for instance, in the world of the Cauchy distribution—networks and other phenomena are not scale-invariant. Miracles are no more common there, but their quality changes from time to time, and the frequency of pseudomiracles, those caused merely by enormous deviations from the average, does not increase over time.

True miracles, those beyond the reach of current science, can become more or less common. As science advances, fewer phenomena are left unexplained. But at the same time, the new horizons opened up by the advance of science might make us more aware of phenomena we hadn't even noticed before. When a problem is solved by science, the solution almost instantly brings new questions to life.

Transcendent miracles are by definition inexplicable within this framework. If such miracles in fact exist, they are unrelated to the distinctions between Mildovia and Wildovia, so it makes no sense to ask whether they are becoming more or less frequent as Wildovia takes over a larger and larger portion of our lives.

Innovation Theft

After all this discussion about Wildovia looming ever larger in our lives, we ought to ask whether such is actually the case. In the previous chapter, I described a few factors that can bring out Wildovian phenomena: the Matthew effect, an increase in complexity, accumulation, and extreme competition. None of these is recent. The very name "Matthew effect" is derived from a two-thousand-year-old piece of writing, and the other three factors have also existed for a very long time. Extreme competition — think Darwinian evolution — has been around much longer than *Homo sapiens*.

At least two of these factors are amplified by the richness of the human imagination: increase in complexity and accumulation. Human fantasy has never shied away from the world of Wildovia, even in areas where the laws of Mildovia obviously dominate. According to the Bible, Methuselah lived for 969 years, and Noah was 600 when he built his ark — and then lived another 350 years. The Deluge itself is Wildovian according to our present knowledge, for while floods might possibly be scale-invariant, there is no evidence that floods like Noah's could have occurred in biblical times. (There is some speculation that the flood myth refers to the creation of the Black Sea, when a rise in sea level after the last Ice Age filled the Mediterranean until it broke through the Hellespont and flooded the basin behind it.) If you live on a moderately high hill, even forty days and forty nights of rain is nothing to fear, yet the biblical flood submerged the three-mile-high summit of Mount Ararat.

Innovation, a product of the human imagination, has always underlain the development of civilized economies. It automatically amplifies the effects of all four factors contributing to Wildovia. Most innovations increase the complexity of the world and the range of available techniques and products. Innovation forms the basis for the accumulation of knowledge and frequently of capital. In economic life, it is the motor of competition. That is all fairly obvious, but it may not be so obvious why innovation also forms the basis of the Matthew effect.

Innovation is generally considered a positive force, which is a reasonable point of view, because innovation leads to the emergence of new, more efficient, more convenient, more useful techniques and products. But every innovation creates losers—a sort of theft by innovation. It always causes someone to lose what had formerly been an advantage. Before the advent of the innovation, there was someone who satisfied a demand that the innovation now satisfies better, more enjoyably, or more efficiently. That is why innovation typically leads to the Matthew effect: "For unto every one that hath shall be given, and he shall have abundance: but from him that hath not shall be taken away even that which he hath." Recall our friend Marina, who came up with the innovation of turning a fishing boat into a tourist attraction. Perhaps her innovation will cause the economy of

the area surrounding Marina's marina to grow, creating winners—
Marina herself, tour guides, perhaps the boat captain—and losers,
namely the fishermen who will lose their jobs.

In our era, innovation has accelerated more than ever before, and
it is therefore no wonder that we are just beginning to feel that the
Wildovian world is playing an increasing role not only in our imaginations, but also in our lives.

Mildovian Models in Wildovia

When Mandelbrot first began appearing at economics conferences in the 1960s, he suggested that economic models based on the
Gaussian distribution should be replaced with models based on the
Cauchy distribution. No one took him seriously. Later, having discovered the scale-invariant world, Mandelbrot softened his views,
and from the 1980s on, he no longer suggested basing economic
models on Cauchy's wild distribution. He would have accepted any
scale-invariant distribution. But the representatives of mainstream
economics (including some who had received a Nobel Prize and some
who soon would) would not even do that. Not until after the 2008
crisis did anyone seriously consider replacing the Gaussian curve—
in some models—with a scale-invariant distribution.

Most economists still shy away from such a radical step. They
believe that scale-invariant distributions do not guarantee that an
economy will have stable states of equilibrium, and if such a model
became the norm, there would be a devastating loss of confidence in
the political and economic systems. Proponents of the model maintain that this position is nothing but intellectual cowardice; if the
world happens to run chaotically, then a chaotic model must be used
to describe it. I agree with this objection, yet it is important to rec-

ognize that the models we use to describe the world have important practical consequences; how a model of a human institution is received can affect how that institution behaves.

Another source of economists' wariness about using scale-invariant models is that while such models might cause economic crises to become less frequent in the long run, they would lead to such a steep increase in the price of options that the farmers and millers we discussed above could no longer afford them. The price of options would swallow up their entire profit and likely more. So they would end up taking risks that were beyond their level of risk tolerance, and when a crisis came, they would not be able to survive it the way a professional investor could. We saw above that investors are psychologically better prepared than farmers and millers to tolerate crashes; below, we shall see how the mechanism of the rich man's junk heap helps them recover from a crash.

In sum, it may be that Mildovian models are more beneficial to the practical operation of an economy, even if they are inferior to Wildovian models in describing how the world actually works. We should continue to use them, just as we continue to design automobiles and household appliances using classical Newtonian mechanics, even after the discovery of relativity and quantum mechanics, because the Newtonian version is better for everyday use. And because the model is useful, it continues to be improved, just as Mildovian models of economics continue to evolve.

The economy is somewhat chaotic even though it is driven largely by the purely Mildovian models used by the vast majority of investors. The paradox is that if the models were chaotic as well, the economy would become much more chaotic than it already is; better models would lead to worse performance. From a practical point of view, we may be better off continuing to use models that promise us a degree of market equilibrium. The Wildovian models tell us that huge economic crises will emerge from time to time, much more frequently than predicted by the Mildovian models. But we are probably better off having to survive the occasional unpredicted crash than letting the economy become even more chaotic than it already is. And not just the economy. As I suggested in the preface, perhaps we are able to have a somewhat civilized society only because we believe we are somewhat civilized. We use a Mildovian model even though we are a Wildovian species. (This is only a hypothesis, but below we will see some compelling arguments for its validity.)

I do not mean to say that theoretical economists should not experiment with wilder models that might describe the world more precisely. That is their intellectual duty. But in everyday practice, it seems worthwhile to stick to the time-tested Mildovian models—even though we will be occasionally taken by surprise by a market collapse—and to try to develop those models in a way that reduces the frequency and devastating effects of such catastrophes.[8]

10

Life in Wildovia

Development and crisis spring from a common root.

If you are not an aficionado of higher mathematics, you will be pleased to learn that I will introduce no more of it for the rest of this book. There is much more that I might have discussed—for example, there are many exciting results concerning the edge of chaos, that fascinating border between Mildovia and Wildovia—but such things can be read about elsewhere.[1] I have presented enough mathematics to back up the points I want to make.

The mathematical concepts that we have been discussing will not help us get rich quick or tell us how to use any enormous amount of money that might fall into our laps. But they will help us get to know the properties of Mildovia and Wildovia so that we can better prepare for at least the first two types of miracles: the pseudomiracles that follow from the very nature of Wildovia and those true miracles that are unexplainable by today's science. We cannot prepare for transcendent miracles. But if such miracles do in fact occur, they are so rare that there may be no point in even trying to prepare for them.

Our friend Taleb, as we have seen, is a serious investor, so he knew precisely how to manage the money he won on September 11, 2001. He stuck to his investment strategy, only now with higher stakes. He continued to buy packages of options that would enable him to win big if the global economy collapsed, and he lost money almost every day until the 2008 crisis made him even wealthier.

Let Us Not Follow Taleb . . .

What would happen if every investor, or at least a large number, followed Taleb's strategy? It is not what most people might think. The answer is not that prices for betting on a total crash of the global economy would skyrocket to the point that such speculation would no longer be worthwhile or feasible. That is the logic of Mildovia, and we are now in Wildovia.

It is true that the price of options that would yield large returns in the case of a crash would rise. But since a significant part of the economy operates in parts of Wildovia, where the standard deviation of large gains and losses is infinitely large, no matter how high the price of such options might rise, they would still be worth having in the event of a major Wildovian crisis. Since the standard deviation of a distribution allows us to determine the likelihood of a rare event, you cannot put a maximum price on something with an infinitely large standard deviation, at least not with the aid of mathematics based on our notion of number. In that sense, the options that Taleb employs will always be underpriced. Nevertheless, I do not recommend such a strategy for most people, since it will be worthwhile only for investors who, like Taleb, are willing and able to lose money day after day for many years.

One reason I don't recommend such a strategy is that it is extremely difficult psychologically. Malcolm Gladwell, in his book *What the Dog Saw,* writes that even Taleb often needed his coworkers to keep him going when his losses, sometimes heavy ones, dragged on and on. Most investors lack Taleb's defining experience of seeing Lebanon, which had flourished for centuries and was considered the Switzerland of the Middle East, collapse overnight and turn Taleb and his family into refugees. Perhaps the shock of that experience was psychologically useful in helping Taleb to follow his investment strategy, since he knew that even if he lost everything, he would not be at his life's lowest point.

A second reason I don't recommend Taleb's strategy is that many people who follow it will slowly "bleed to death" before the next major crisis. In chapter 1, I quoted Taleb saying that he would not go bankrupt but only bleed to death if there was no market collapse for a long time. The more people who follow his strategy, the greater the collapse will have to be to counter all the losses incurred in waiting for it. With our knowledge of Wildovia, we can be certain that sooner or later, an arbitrarily huge crash will occur, but will we still be alive? And will we have spent years impoverishing ourselves waiting for the big win that never comes?

The famous British economist John Maynard Keynes did not think much of investment strategies envisioned over a very long term. He thought that in the long run, unquantifiable economic and political uncertainties overwhelm the quantifiable risks. Nevertheless, he was a very successful investor, amassing a large personal fortune and substantially outperforming the market indices during the quarter century that he managed the endowment of King's College, Cambridge. He believed that it is important to consider short-term problems and not just say that in the long run, things will turn out all right. "*In the long run,*" he famously wrote, "we are all dead."[2] This conclusion is something we need to consider before adopting Taleb's strategy. Although a crash will always be underpriced in Wildovia, the more people who follow Taleb's strategy (or something like it), the higher the stakes will become, because the price of options in favor of such strategies will rise, and the minor and medium-sized crises that have made money for Taleb will not be able to compensate for the large daily losses of his imitators.

One way to lessen the danger of bleeding to death is to begin to lower the risk level of our investments after a time. That will also, of course, reduce our future profits when the really big crisis eventually occurs, but it will also delay the day when we go belly up because no market crash occurred. Yet this strategy has the downside that bleeding to death more slowly is even harder psychologically than sustaining continuous losses, though it is mitigated by the fact that we are in total control.

The third reason not to follow Taleb is the most important of all. It does not have to do with one person's economic and psychological well-being but with the economic welfare and cohesion of the entire society.

Taleb is not interested in building a flourishing society. The only thing he cares about is making a killing when a crisis hits, and his only worry is that he might run out of money before anything really bad happens to the economy. He bides his time, hoping that things will go really badly sooner rather than later. Meanwhile, he is doing nothing to contribute to the useful junk heap of society, which is built by ideas, insights, and discoveries made by those who are trying to make things go well. These are the things that create the treasures hidden in our attic, even if they fail in their first incarnation. Taleb's strategy, on the other hand, doesn't build anything. It may even contribute to the emergence of a crisis as a self-fulfilling prophecy. That is why the idea of banning or at least strictly regulating such strategies is resurrected from time to time.

The history of the past few thousand years shows that there have always been economic crises, both great and small. Knowing what we do about Wildovia, we can be quite certain that the future will also be punctuated by crises. And the effect will only increase as the scale of scale-invariant events expands. But humanity has produced a mechanism that makes it possible to recover from even the most devastating crisis. I hinted at this mechanism above, when I noted that there is more of value to be found in a rich man's junk heap than in all of a poor man's possessions. We will soon see what kind of a regeneration mechanism this principle has created. For now, we simply note that if most investors followed Taleb's strategy, then the time-honored natural mechanism of building on the rich man's junk heap, which has been the basis of recovery for millennia, would cease to function. This wouldn't necessarily be a problem if Taleb's strategy were to substitute a different mechanism that worked better—but it doesn't.

. . . Yet Take Wildovian Phenomena into Consideration

We continue to live mostly in Mildovia, so unlike Taleb, I do not believe in abandoning Mildovian models wholesale and substituting Wildovian ones, even if Wildovian phenomena are now recognized as playing important roles in more and more aspects of our lives. Significant Wildovian events may occur years apart, and in the meantime, we still have to live in Mildovia. Applying Wildovian models would make everyday life a positive nuisance unless one possessed the attitude and life-defining experiences of a Taleb.

But when we think about our distant future, we must recognize that sooner or later, Wildovian phenomena will pop up. Then we will have to make use of the concepts that proved especially useful under Wildovian conditions, but we must do so in a way that will not, insofar as is possible, completely upset our everyday Mildovian lives.

This seems impossible on the face of it, since Mildovia and Wild-ovia are fundamentally different. The following are some important differences, supported by Mildovian models based on centuries of results and by what we have learned of Wildovian models:[3]

- In Mildovia, we should not count on major surprises, since black swans are extremely rare. In Wildovia, we bump into them all the time, and the fact that most of them are only gray swans, because the laws of Wildovia guarantee their occurrence, does not change this. Both the nature and timing of gray swans are unpredictable.
- In Mildovia, it is generally advisable to take medium-sized risks, avoiding both those that are too small and those that are too large; there is next to nothing to win with very small risks, and there is generally too much to lose with really large ones. Conversely, Taleb has demonstrated mathematically that in Wildovia, it is worth taking many really small risks and a few really large ones, while medium-sized risks are to be avoided.
- In Mildovia, history flows slowly, and revolutionary changes occur very rarely. In Wildovia, history happens in huge leaps, and we should count on several revolutions in our lifetime.
- In Mildovia, the most useful behavior in the long run is conformity —that is, following the rules. This applies even to extraordinary talents, although occasionally such individuals discover something that forces the domain of the rules' validity to change somewhat. Very rarely, new (though still basically Mildovian) rules are added to the existing ones, incorporating new instances of "extremities of Mildovia." But in Wildovia, what most often proves effective is nonconformity, the reasonable transgression of Mildovia's rules.[4]

Life in Both Worlds at Once

At some point in my course on economic psychology, one of my students will ask how we know when we have passed from Mildovia into Wildovia. I used to give a purely technical answer: you can recognize Wildovia when the distribution of the phenomenon under investigation resembles a Cauchy distribution rather than a Gaussian one. Sometimes, while remaining technically correct, I would try to be more poetic: you know you are in Wildovia if nonconformity seems like a promising strategy. My students did not seem satisfied with these answers, and I began to feel that they were missing the point. It took me a long time to realize that the correct answer to their perfectly reasonable-sounding question was that the question was ill posed.

We can never tell when we have left Mildovia and entered Wildovia, because it is simply not true that we live in one world but sometimes find ourselves in the other. Mildovia and Wildovia are always present at the same time. Certain things follow the laws of Mildovia, while others follow the laws of Wildovia. In fact, as we have seen, Wildovia contains scale-invariant phenomena that are only mildly Wildovian (gray swans of sorts), as well as much wilder phenomena. In short, there is more than one law in Wildovia.

If we were to insist that the laws of Mildovia comprise all the laws of nature, then we would have to consider everything that contradicts those laws a miracle. Such insistence would make every stock market graph a miracle, since anything that is scale-invariant does not conform to Mildovian laws. It would make every human brain a miracle, since our brains have fractal-like features. Miracles would become a commonplace, and if science's only response were to throw up its hands in despair and assert that miracles are none of its business, then shame on science. Science has rightly expanded its horizons to encompass the laws of Wildovia.

The science of Mildovia continues to increase the scope of what it is able to describe. It is able to offer better models, in Mildovian terms, of phenomena that are by nature Wildovian. For instance, one might try to fatten up the tail of a normal distribution by characterizing the really extreme cases with a separate normal distribution that is appropriately adjusted by placing it several standard deviations from the mean.[5] In this way, one may use normal distributions to model a phenomenon five or six standard deviations away from the mean instead of only three or four, thereby encompassing a few cases that were previously considered true miracles according to our taxonomy.

But if the phenomenon in question in fact follows the Cauchy distribution, then sometimes events will occur that are ten or even a million times as miraculous as what would be considered a miracle in Mildovia. We know this from the example of Phoebe the markswoman, who occasionally stops spinning almost exactly parallel to the wall and then shoots a distance that seems miraculously far. If an economic phenomenon is modeled well by the Cauchy distribution, then trying to squeeze and stretch it into a procrustean Mildovian bed ends up obscuring the entire point of the phenomenon — that is, the fact that there is no standard deviation. Such Mildovian models end up putting a price on the unpriceable, something both Taleb and Mandelbrot would say is rubbish. Those gentlemen have a point, since the science of Wildovia has shown that certain things work according to non-Mildovian laws. On the other hand, we have seen that *not* putting a price on the unpriceable is not an acceptable solution, because then the farmer and the miller would find it impossible to run their businesses efficiently.

Like Taleb, we should be concerned about the possibility of crises in the form of sudden crashes and find ways to prepare for them. We are learning more and more about how, within the framework of Mildovian life, we can prepare for times when we find ourselves at the extremities of Wildovia. We should therefore feel free to continue to improve and apply the methods of Mildovia to our everyday lives, knowing we shall have to live with the occasional catastrophe. But that is better than throwing away our beloved Mildovian lives and giving ourselves over completely to the extremities of Wildovia. The science of Wildovia also has its uses. It helps us appreciate that our Mildovian science inevitably has its limits, and we will bump up against them from time to time.

Recall the Arrow-Debreu theorem, which guarantees that a Mildovian economic equilibrium can exist under certain conditions. But those conditions can be broken by an extreme event. One of the conditions, for instance, is that there be no monopolies, but monopolies form regularly even in Mildovia, since the Matthew effect also works there. Monopolies form in Mildovia for the same reason that geniuses are born: a monopoly, like a single Einstein, has zero standard deviation.

If it is economic equilibrium that we want, we should try to regulate the economy in a way that prevents monopolies from forming, and we should also maintain all the other conditions of the Arrow-Debreu theorem. But the science of Wildovia guarantees that even if we do so, monopolies will occasionally form. We must learn to live with such threats to equilibrium. In chapter 3, "The Source of Miracles," we saw that in every formal (i.e., Mildovian) legal system, loopholes necessarily exist. No matter how we try to regulate the world so that the laws of Mildovia will prevail, Wildovian phenomena will continue to arise. There is nothing to do but live in Mildovia and hope we will be able to recover from the inevitable next crash.

We may have one additional chance to avoid future catastrophes. Perhaps a theory will someday be created that incorporates the laws of both Mildovia and Wildovia, contradictory though they may appear to us. Something like this happened in the history of physics when Newton successfully unified the laws of terrestrial and celestial mechanics. Perhaps the foremost problem in today's physics is that separate and incompatible theories exist for gravitation and quantum mechanics. Each theory applies only to its own domain and is incorrect when applied to the other. Physicists are still searching for a grand unification that will embrace both theories without opening up contradictions.[6] So far, that theory has remained elusive, although most physicists believe they will find it sooner or later. How could it be that three types of interaction (electromagnetic, weak, and strong) can be incorporated neatly into a beautiful and consistent theory while the fourth, gravitation, is off by itself?

In fact, physicists should be happy if the situation turns out to be simply Gödelian—that is, if the two theories just happen to be independent of each other. We should have no problem imagining that gravity is in fact independent of the system formed by the other three interactions, and if that is the case, there will never be a grand unified theory. I find it likely that something similar applies to Mildovia and Wildovia—that is, that they are mutually independent and that it is impossible to create a scientific theory that encompasses both worlds.

Wildovian phenomena show up from time to time in Mildovia, since there are some things, such as innovations, that always work according to Wildovian laws, even in Mildovia. They are described by different mathematics and different logic from those that describe the phenomena of Mildovia. But the difficulties are mitigated by the fact that both types of logic can be taught and learned within the Mildovian framework, in Mildovian schools, just as the thoughts of a genius can be made comprehensible by not even remotely genius teachers for the masses of those who are not geniuses.

The Nature of Thinking Both Ways

Most people prefer to live by the rules of Mildovia and deal with Wildovia only when necessary. But most of us will at one time or another be confronted with a Wildovian event—for example, if we are among the unlucky ones who lose our livelihoods to some significant innovation. In that case, we just have to find some other way to earn a living. We can prepare for this eventuality even if we can't imagine how the situation might actually arise. And trying to predict when such an event will take place is mathematically impossible. We must learn to think in Mildovian and Wildovian terms simultaneously, which means that we must be somewhat familiar with the laws of both worlds so that we may calmly live our familiar Mildovian lives while being prepared to react to a Wildovian event at any moment.

This kind of "thinking both ways" is not precisely what George Orwell meant by "doublethink" in his novel *Nineteen Eighty-Four,* though to the extent that we are talking about the "power of holding two contradictory beliefs in one's mind simultaneously, and accepting both of them," they are more or less the same thing.[7] But there is an important difference. In Orwell's novel, doublethink explicitly concerns the lies that serve the Party's current interests and must be believed unconditionally: "To know and not to know, to be conscious of complete truthfulness while telling carefully constructed lies, to hold simultaneously two opinions which cancelled out, knowing them to be contradictory and believing in both of them, to use logic against logic."[8]

Thinking in both Mildovian and Wildovian terms serves no political purpose. It simply gives us a better understanding of the world around us. Mildovia and Wildovia do not stand in opposition. They exist side by side. According to Orwell, "Even to understand the word *doublethink* involved the use of doublethink."[9] There is nothing so sinister going on here; to apply both kinds of thinking, it is enough to understand the rudiments of Gödelian thought. And if we are accused of contradicting ourselves, we can agree with Ralph Waldo Emerson that "A foolish consistency is the hobgoblin of little minds" or say along with Walt Whitman, "Do I contradict myself? Very well, then I contradict myself (I am large, I contain multitudes)."

The world may contradict itself too. The domain of validity of every model is bounded, and if some aspect of the world is outside that boundary, then understanding it requires a different model. Bearing this in mind, we are free to live our more or less Mildovian lives, no longer naively, but knowing that small, large, or even enormous crises sometimes occur.

There will always be miracles, pseudo and real, positive and negative. Development and crisis spring from a common source. We must use the periods of development to prepare to withstand crises and recover from the devastation they cause. We have to save while we can for when the next crisis hits. Not that we should count on our accumulated wealth being able to see us through the crisis, for there may always be a crisis so big that it will bankrupt us. What we need is a rich enough junk heap to begin the postcrisis recovery.

The Rich Man's Junk Heap as a Constructive Mechanism

Here is the context of John Maynard Keynes's famous line, "In the long run we are all dead." He goes on to say, "Economists set themselves too easy, too useless a task, if in tempestuous seasons they can only tell us, that when the storm is long past, the ocean is flat again."[10] Keynes was interested in how to sail rough waters. And our subject is how to prepare for those rough seas when the ocean is calm, to prepare for a time when the great storm—which must come sooner or later—has subsided and the ocean is flat again but our ships have been lost. We want our junk heap to be rich enough that after all the havoc wreaked by the storm, we have something of value left over.

It may seem hard to believe that the main mechanism for recovery is a heap of junk. After all, things enter a junk heap precisely because we believe they have no value. They do have no value if things are going well, but they regain their value as soon as things start going badly. Such things are not at all rare in nature. For instance, there is stuff that seems like cellular junk floating around in the human bloodstream. These are fragments of protoplasm called platelets, or thrombocytes, that just hang around with seemingly nothing to do as long as things are going well. But when a crisis appears in the form of a blood vessel injury, the platelets rush to the scene to help fill the tear in the vessel.

Taleb's main message seems to be that we should avoid the losses caused by devastating storms. But crises are inevitable, and when they strike, they are frequently catastrophic for large numbers of people. So on the level of society, Taleb's recipe is useless. It is also highly morally questionable. I think Columbus and Magellan would have despised someone who said, "Don't sail; it is too dangerous. Stay home and bet against the suckers who are foolish enough to venture out on the high seas."

The point, therefore, is not to try to avoid loss when a crisis hits. It is to make proper use of the positive miracles of development during periods of calm. As we saw in chapter 9, "The Levels of Wildness," some people are robbed, so to speak, by innovation. They once satisfied a demand that is now better satisfied by the innovation, so they suffer some degree of loss. Methods, products, and technologies that once worked get thrown onto the junk heap of history. But after a crisis, these methods and technologies may prove to be still useful—indeed they may save our lives. We have to use the periods of development to build as rich a junk heap as possible to help us recover after a crisis so that development can resume. The richer the junk heap, the better the chance of recovery.

PART FOUR
Preparing for the Inconceivable

We don't know what the future holds, but it is
up to us to help determine what it can be.

11

Adapting to Wildovia

It is not only the worlds of art and fashion that turn to the past for new ideas; long-abandoned ways of doing things can help solve a societal crisis.

When I introduced scale-invariance, I noted that the recent increase in Wildovian phenomena might have four causes: an increase in complexity, intense competition (e.g., Darwinian evolution), the Matthew effect, and accumulation (of knowledge as well as capital). If we are looking for principles to help us survive a Wildovian regime, we should first of all consider these four factors and how we might be able to ameliorate their effects or even turn them to our advantage when we see that they are leading to a Wildovian state of affairs.

Intelligent Simplicity

An increase in complexity may contribute to the emergence of Wildovia, so in preparing for the Wildovian environment, we should figure that the value of simplicity—at least of *intelligent* simplicity—will increase. Since it is the imagination that rules in Wildovia, it is not surprising that this important principle has been understood for a long time and is frequently realized in various branches of the arts. I offer as an example one of my favorite sculptures, the zero-kilometer stone in Budapest (figure 22). This sculpture by Miklós Borsos, erected in 1975 at the Buda abutment of the Széchenyi Chain Bridge, is the reference point from which distances along all the country's main roads are counted. This location has long been the zero reference point. Other sculptures were there previously, but they were much more complex, just as the reference point for measuring road distances in the Roman Empire, the Milliarium Aureum (or Golden Milestone), in the Roman forum, was also more complicated (perhaps because the Romans did not possess the simple concept of the number zero). Miklós Borsos realized this abstract concept of zero in a way that is simplicity itself: a nicely formed zero. Most men see it as a female symbol, while most women see it as a male symbol, which shows that it serves as a point of departure for more than just highways.

Figure 22. The zero-kilometer stone in Budapest (photograph by Vera Mérő).

Like this sculpture, a person can also be intelligent and simple at the same time. One of my former students excelled in performing a challenging yet extremely monotonous task incredibly rapidly. It required sorting a collection of sentences according to opposing categories, such as good-bad, severe-permissive, and frugal-profligate. Because the task was monotonous, you needed a certain simplicity to keep at it, but you also needed a degree of intelligence to do it accurately. All the other students eventually gave up; the more persistent continued for a while, but they worked ever more slowly, and even then they didn't do a good job, primarily because they thought too much about what they were doing. It turned out that a successful strategy was to sort the sentences based on a first hunch and not to think too much. Yet there was this one student who worked cheerfully and rapidly and, as subsequent analysis proved, in an intelligent and logical way.

So I decided, as a little bonus, to pay this student to take a Mensa qualifying test, a sort of high-end IQ test that separates the top 2 percent of people from the other 98 percent. I thought she needed intellectual confidence and that the test would provide it. She was reluctant to take the test, saying it was only for smart people, not for the likes of her. "Let's give it a shot," I said, and downloaded a test similar to the official one. She fell short of qualifying by only a single point. "There, you see?" she said. "I told you that I was not one of those supersmart people." I almost exploded. She had tried something for the first time in her life, came up one point short, and was ready to give the whole thing up. For sure, I thought, she was going to pass the next time.

I downloaded an even harder test. She completed it flawlessly within the allotted time, which was better than I could do. "How did you get the correct answer to the last question, the one with the geometric figure?" I asked. "Don't you see? Zap from the right, zap from the left, and then bang," she replied, with an accompanying hand motion. Then I saw it. The solution was to take the symmetric difference between the two sides of the figure divided vertically through the center and then rotate each column upward by as many steps as the number of the column. The subsequent analysis of why the solution worked was by no means obvious.

I concluded that this young woman possessed intelligence of a very high order (she solved the Mensa application test perfectly) that at the same time was extremely simple in that she didn't bog down her thinking with complex concepts. However the tides of Wildovia might ebb and flow, she will find a place for herself, for she can realize what Taleb suggests as a strategy for Wildovia: to be "a fool in the right places."[1]

In a sense, Rubik's cube also realizes this advice. The wonderful thing about its mechanics is that no matter how you twist and turn it, its internal structure remains exactly the same, even though, of course, the colors on the faces get mixed up. Ernő Rubik first tried to realize the cube using a complex arrangement of rubber bands, but the bands became so tangled that after a while, not even the strongest player could twist the cube. He finally came up with a solution that, though far from obvious, nevertheless worked in an extremely simple way. The miracle of Rubik's cube as a mechanical structure arose from the almost foolish simplicity of the way in which it operates.

The principle of scale-invariance employs a similar simplicity. It allows us to identify the areas within Wildovia that may be considered mildly wild. It is a very simple concept that something should have the same structure on a small scale as on a large scale, yet as we have seen, the self-similar object known as a fractal can be remarkably (even infinitely) complex. Scale-invariance has been successfully exploited only in the applied arts such as computer graphics. No one has figured out how to harness the principle on a large technological or industrial scale.

Comparative Advantages

Another important component in the emergence of Wildovia is extreme competition. Darwinian evolution, for instance, operates not only in the natural world, but also in economics. Evolution suggests a general principle—known for almost two hundred years—that works in both Mildovia and Wildovia but that can be especially useful in the latter.

Darwin's theory of natural selection is often misunderstood. In the popular imagination, "survival of the fittest" has come to mean the triumph of the strong and extinction of the weak. But that badly misses the point. Natural selection might seem cruel—"Nature, red in tooth and claw," as Tennyson put it—yet natural selection permits forms of life to survive that would be vanquished in single combat.

For simplicity's sake, let us consider a hypothetical example. Suppose in a school of fish there are certain individuals that are relatively weak and unable to compete successfully for food against their stronger fellows. These fish will not do very well. They will have relatively shorter lives and mate less often. But suppose a few of these weaker fish find themselves washed into some murky backwater where the food is scarce and hard to get, yet there is no competition because the stronger fish prefer to fight it out on the high seas where food is abundant. Some of our scrawny little fish may stay there and survive, even if they are in every way inferior to their stronger cousins. Our fish have found themselves a Darwinian niche, and over time, separated from their seagoing brethren, they may form a separate species, and this new species may go on to develop traits, such as vision or balance or maneuverability, that are superior to those of the species that they left behind.

A similar principle in economics was discovered in 1817 by David Ricardo, more than forty years before Darwin's *Origin of Species* was published. Ricardo, a British economist, was analyzing international trade, and he discovered that two countries may agree on a mutually favorable deal between them even if one country is able to produce everything in the agreement more cost-effectively than the other one.[2]

Let us suppose that countries X and Y—let us call them Existan and Wyistan—are producers and trading partners in food and clothing and that each country can consume as much of each as it can either produce or acquire in trade. Existan can produce either three times as much clothing or two times as much food a day as Wyistan, depending on whether it prioritizes the production of clothing or food. If Wyistan is able to produce more food than it needs to feed itself, then it is better off if after having produced its daily food requirement, it does not switch to producing clothing but continues to produce food. Existan will buy Wyistan's food surplus in exchange for more clothing than Wyistan could produce for itself in the same amount of time. This arrangement is also good for Existan if the exchange rate between clothing and food allows it to exchange a quantity of clothing for a larger quantity of food.

Although Existan produces both items more efficiently than Wyistan, it actually has a disadvantage in producing food compared to Wyistan. This strange observation is the point of Ricardo's discovery, which can be best understood if we consider how much clothing production each country has to give up in order to produce one unit of food. Existan sacrifices one and a half units of clothing to produce one unit of food, while Wyistan sacrifices only one unit. In Ricardo's phrasing, Wyistan has a *comparative advantage* in producing food, even if it is able to produce a smaller absolute quantity of food per day than Existan. Ricardo discovered that this principle generally holds in international trade.

The logic of comparative advantages is exactly the same as that of Darwinian evolution. Wyistan and the weak fish have a comparative advantage not because they are better at producing food or foraging in the backwaters but because their performance in everything else is much worse. It is less of a sacrifice for the weak fish to hunt for food in the backwaters than it is for the stronger fish, so the weak fish are better off living in those backwaters than trying to compete with others of their species. When Wyistan produces food, it makes less of a sacrifice in the production of clothing than Existan would make in producing food. Paradoxically, it is their inadequacies that enable their survival, provided they can find a niche in which they are relatively less inadequate than their competitors.

Neither Darwinian nor economic competition is an all-or-nothing affair, and despite their merciless aspects, both types of competition assist in the formation and survival of a variety of life forms and economies. Those that can find a comparative advantage have a chance of surviving, even if they are at a disadvantage in everything else.

Comparative advantage became a fundamental principle of international trade, and this is how it continues to be interpreted and taught in business schools. But it is much more general and even extends down to the level of the individual. In my book *The Biology of Money*, I illustrated this concept through the example of a lawyer. Imagine a lawyer—let's call him Remington Underwood—whose hobby is typing and who in fact is so good that he has just won the bronze medal in the world-championship speed-typing competition. Now, should Mr. Underwood type his own briefs at his law firm? Probably not. Suppose Underwood's lawyerly work is worth $200 an hour to the firm, and suppose that he can type twice as fast as the best of the office typists, who earn $35 per hour. Each hour that he devotes to typing a brief saves the firm $70 that it would otherwise have to pay a typist for two hours of work, but it costs the firm $200 in lost productivity. Underwood's boss is justified in telling him to confine his typing to his free time unless he wants to give up being a lawyer, take a large salary cut, and become a typist. It is no surprise that Remington Underwood decides to keep practicing law and give up typing on company time. This is true even if he is one of the world's best typists yet only a so-so lawyer. His comparative advantage is still practicing law.

Underwood has another great advantage in his decision to keep lawyering. As a merely good-enough lawyer—say, the hundredth best at his firm—he will likely not feel that he is in intense competition with his colleagues, certainly not the level of competition he experiences in keeping near the top of the international speed-typing rankings. So he likely doesn't feel he has to break his neck doing legal work. That should leave him sufficient spare time for his hobby of typing. Should the firm hire three brilliant young lawyers, Remington Underwood will fall to 103rd best, but that shouldn't make a big difference. However, if three brilliant young speed typists hit the international typing circuit, Underwood will drop from third place to sixth, a change that might make him lose interest in typing altogether.

This does not mean that Remington Underwood can slack off as a lawyer. If his productivity falls so that he is no longer worth $200 an hour to his firm, his superiors might not mind if he types his own briefs, and if his profitability slips to $70 an hour, he might find himself being asked to type other lawyers' briefs. At that point, his absolute advantage as a typist will finally have become a comparative advantage as well.

We do not have to be the very best at something. There are many of us and room for only one top dog. Someone who is the best in the world at something might end up doing something else because his or her comparative advantage lies elsewhere. What is important is to find a field of endeavor in which we have a comparative advantage relative to others. It is not always easy for those who can do a number of things well to make the optimal choice. But it is worth making the effort to find a satisfactory field, not only because it increases our sense of well-being, but because we may also decrease the stress caused by extreme competition. Once we have found a field in which we can maximize our relative advantage, we may not have to work ourselves to death, leaving us more time and energy to pursue things that interest us.

Let me emphasize this one more time, since I have learned from experience that students have difficulty understanding this concept. Our comparative advantage does not generally come from what we are good at. It depends on what other people are good at and even more on how weak we are in the most popular fields. If we are much weaker than everyone else in every field save one, where we are only a little weaker, then that one field is our comparative advantage, like Wyistan's advantage in food production.

It may seem paradoxical, but it is often easier for those who are relatively weak to find their comparative advantage, and so have an optimal chance for survival and success, than it is for those with many strengths. Those who are strong in many fields tend to choose from the most popular ones, and there they are likely to find themselves confronted with someone even stronger. They might fail without ever understanding how so many people who were much weaker in every way have managed to flourish. If we learn to think in terms of comparative advantage, we are more likely to find ways of survival when we find ourselves in Wildovian situations—that is, in times of great change.

The Matthew Effect

The third component in the emergence of Wildovia is the Matthew effect. As we have seen, the fact that the rich tend to get richer and the poor poorer works automatically in some cases and can lead to Wildovia all on its own. But the Matthew effect is much more common than just those cases in which it follows mathematically from the theorem of Barabási and Albert. It pops up in many situations in which the conditions of that theorem are not met exactly, and it can be brought about by almost unnoticeably small factors.

In his book *Outliers*, Malcolm Gladwell presents one such phenomenon among ice hockey players. He lists Canadian pro league hockey players with their respective dates of birth. The year didn't matter; he was interested in the month, and there he found something truly weird. It turned out that approximately 40 percent of the players were born in the first three months of the year, 30 percent in the second quarter, 20 percent in the third, and only 10 percent in the fourth. Only a few of those final 10 percent were born in December.[3]

The cause of this remarkable phenomenon is not some astrological configuration that bestows special ice hockey talent on those born under the sign of Aquarius or Pisces. It is that many Canadian children are introduced to ice hockey at a very young age. They learn to skate almost before they can walk. The most talented skaters are

selected for further training based on their performance at competitions in which children born in the same calendar year compete
against each other. At such a young age, a gap of a few months makes
a great deal of difference, so children born at the beginning of the
year have an advantage over those born at the end. Here is where
the Matthew effect kicks in, to the point that the early advantage
of the older children dramatically increases their chances of making
it into the adult professional league, while those born in December
are permanently left behind. The older, more developed children in
each age group are seen as more talented; they receive more playing
time and more training from coaches and get more encouragement
to improve their skills, advantages that get them more playing time
and more attention. *For unto every one that hath shall be given, and he
shall have abundance: but from him that hath not shall be taken away
even that which he hath.*

Reading about this odd example, someone else had his curiosity
piqued and checked the birth months of the national soccer team
of the Czech Republic, where the qualification system is similar to
that in Canadian ice hockey, and found the exact same result. Then
someone looked into Major League Baseball and discovered an un-
expectedly large number of players born in the third quarter, and lo
and behold, in baseball, the cutoff birth month for training camps is
July rather than January.

Of course, it is not enough to have been born in January to succeed at ice hockey. You also have to have talent and intrepidity. But
because of an artifact of the qualification system, if you were born in
December, you have to have even more talent than if you were born
in January.

The Matthew effect was first described scientifically in 1968 by
the American sociologist Robert K. Merton (father of the Nobel laureate economist Robert C. Merton, whom we met in chapter 6 as

the third author of the Black-Scholes formula). He is also the person who gave it its biblical name.[4] Merton looked at the recipients of scientific awards and prizes and discovered that if someone was honored at a young age, the award often worked as a self-fulfilling prophecy, leading to more citations and further awards at a rate that was more or less independent of the scientist's subsequent accomplishments.

At around the time the Matthew effect was discovered, social psychologists discovered a similar phenomenon, which they termed the Pygmalion effect. The American researchers Robert Rosenthal and Lenore Jacobson received permission to administer an IQ test at a school. The pupils' teachers were told that this was not a regular IQ test but the "Harvard Test of Inflected Acquisition" and that it would indicate which students were likely to exhibit a significant increase in academic performance in the coming year. After the test was administered, the teachers were given a list of their students who had scored in the top 20 percent and were therefore ready to "blossom." The teachers were not told that the scores had been assigned at random, irrespective of students' performances on the test. At the end of the year, the test was given again, and among the youngest children, those in grades 1 and 2, those students who had been labeled "ready to bloom" showed greater average improvement in their test scores than their peers.[5]

This experiment shows that the Matthew effect also works in Mildovia, and we have seen previously that it can lead to Wildovian situations. But there is a side effect that enables many people, such as our friends the farmer and the miller, to live their quiet Mildovian lives and yet survive when Wildovian circumstances bring turbulence to the river of life.

As a result of the Matthew effect, not only will all unto whom it is given increase their own wealth more and more, but their junk heaps will also become richer and richer as they consider ever more valuable things disposable. If, for instance, a craftswoman is favored by the Matthew effect, she may decide to buy better tools and more expensive materials. Her old tools and materials are not worn-out trash; they are perfectly good items that have been replaced by better ones. And since our craftswoman, typically of her kind, never throws anything away that might someday come in handy, she puts her old tools and materials up in the attic along with other leftover items. And there they remain until a crisis comes along, when some of those items might prove useful. A solution to almost every problem can be found in a good craftsperson's attic. It might be a makeshift solution, but it is often good enough to keep things going.

So every one that hath and unto whom it is given, such as a rich man or a good craftsman, can take greater risks than others because he has learned that he can usually find a way out of a critical situation. He will undertake tasks that look unsolvable, figuring that he will come up with something, and with the aid of his attic, there is a good chance that he will. After a market crash, a good investor can generally find something in his junk heap to help him pick himself up and start all over. In any case, he has a better chance of finding it than someone with a less-well-stocked junk heap or attic. Moreover, both the craftsman and the investor have experience creating something from scratch, and that experience will prove valuable in difficult situations.

Thus the Matthew effect does not, as Merton thought, merely describe the emergence of social injustice. It also explains a socially useful phenomenon, the well-endowed junk heap. The junk heap provides a cushion for individuals who are willing to take extraordinarily large risks, not out of necessity, but willingly, because those risks are worth the unavoidable failure that comes with the territory, just as every boxer has to live with regularly taking it on the chin.

Accumulating the Well-Endowed Junk Heap

The fourth component for the emergence of Wildovia is the accumulation of capital and knowledge. The Matthew effect works not only on the level of the individual, but also on the level of society. Unto every society that hath great knowledge and capital shall be given, but from a society that hath not shall be taken even that which it hath. For example, the most talented and energetic people leave the poorest countries to work abroad. John von Neumann could hardly have built a computer in Hungary, and Eugene Wigner could not have designed a nuclear reactor there. In poorer countries, the most attractive local opportunities are generally exploited by foreign investors, often to the benefit of the poorer society, since it does not have sufficient capital of its own to seize the opportunity, but such investors also send much wealth abroad to richer countries.

As a society accumulates capital, its money and time can be devoted to pursuits beyond what is essential for survival, and things can be done solely for the sake of enjoyment, making life richer and more meaningful. Out of those pursuits may arise a radically new, socially useful miracle. Consider the example of Michael Faraday, discoverer of the esoteric physical phenomenon of electromagnetic induction (in 1831) and of the laws of electrolysis (in 1833). In 1850, William Gladstone, then the British chancellor of the exchequer, is supposed to have asked Faraday whether electromagnetism had any practical use. Faraday is supposed to have replied, "Why, sir, there is every probability that you will soon be able to tax it."[6] And indeed, Faraday's discoveries made possible the electrical infrastructure we have today.

It is not just the rich people's junk heaps that get larger and more valuable. Another result of accumulation is the increase in the junk heap of society as a whole. There is an accumulation of societal knowledge to fall back on should current ways of doing things fail. It is not only the worlds of art and fashion that turn to the past for new ideas; long-abandoned ways of doing things can help solve many crises. For example, after the 2008 economic crisis, certain kinds of options trading were regulated in a way much like what had been done in 1720 (when Newton lost his fortune).

A great deal of effort in both the business sector and the sciences goes into experimenting with ideas, solutions, and models to discover what works and what doesn't. All the while, the junk heap of the future is being built, as today's state-of-the-art technology becomes tomorrow's obsolete junk. Perhaps instead of a junk heap, we should call it a museum of science, industry, and technology, in which you can find all sorts of things that might come in handy. This rich junk heap gives both individuals and society a chance to recover after a crisis. This fact is reflected in Todd Buchholz's successful book *New Ideas from Dead Economists*. After the crisis of 2008, several countries successfully reached back to economic methods, such as counter-cyclical economic policies, that had been considered obsolete.[7]

Analyzing the four components that lead to Wildovia has helped us to identify some tools that can be particularly useful for achieving success when a Wildovian menace strikes. Two additional principles useful in Mildovia are absolutely unavoidable when we cross over into Wildovia: antifragility and convertible knowledge.

12

Antifragility

If the winner doesn't take a few on the chin as well, it's not a fight; it's a massacre.

Nassim Taleb has always been interested in what can be done against the fragility of things. He wrote a book on just this topic a few years after *The Black Swan*. It is called *Antifragile*. He chose as his title a word that does not exist in English to emphasize that the subject of the book was a new concept that required a new name. I think this discovery of Taleb's is sufficiently important in the set of items necessary for surviving in Wildovia that it deserves its own chapter.

What Is the Opposite of "Fragile"?

If you ask your friends and acquaintances what the opposite of "fragile" is, you will get answers like "stable," "massive," "robust," "sturdy," "vigorous," "rugged," "unbreakable." But these antonyms go only halfway, indicating only that if something is not fragile, it resists being broken. Why not describe something that glories in any attempt at breakage? After all, the opposite of "good" is "evil" and not some indifferent state; the opposite of "positive" is "negative" and not "zero"; the opposite of "destruction" is "construction," not maintaining the status quo. But the opposite of "fragile" is, apparently, nothing better than "unbreakable." We need some other word, say "antifragile," to describe something that not only resists being broken but also benefits from any attempt to break it. That is the concept Taleb discovered.

It is our everyday experience that provides us with the notion of opposites. The opposite of "big" is "small" and not "average-sized," not even "zero-sized," and certainly not "negative-sized." It can also happen that a concept has more than one opposite. Thus the opposite of "infatuation" can be "indifference" but also "detestation." The opposite of "inflexible" is usually "flexible," but the opposite of "flexible" can be "fragile" as well as "inflexible," which suggests that an opposite of "inflexible" might be "antifragile." And indeed, a spring's flexibility encompasses more than just its not being harmed by being compressed; it positively benefits from compression, for that is what gives it the ability to spring back. Nonetheless, "flexible" is not the opposite of "fragile."

No word I know of in any language directly expresses a benefit to something as a result of an attempt to break it. But such things do exist, and not only in the special case of the spring. Taleb's nice example is the hydra from Greek mythology: it possessed many heads, and if one of them was cut off, two new ones grew in its place.[1]

The hydra is a mythological creature, but its regenerative ability is common in the biosphere. Bacteria, for example, become more resistant when we attack them with antibiotics. The main driving force of biological evolution is precisely that competition and stressful environments weed out the less adaptable, to the benefit of those that can profit from the otherwise adverse environment. And the notion of antifragility applies not only to species, but also to individual organisms. Thus muscles and bones develop by being exercised—that is, by being stressed—while pruning plants stimulates their growth. And something like this occurs in human thought: we sometimes need to be pushed to think harder. A mathematics teacher of mine used to say that some students are like certain medicines: shake before use. He would do it, too, and not only figuratively.

Psychologists use the term "resilience," which describes a quality of a person who becomes stronger as the result of facing life's challenges. This is summed up in a line from the 1960 hit musical *The Fantasticks:* "Without a hurt, the heart is hollow." Thus "resilience" describes a special case of antifragility, but it remains confined to psychological jargon. The technical term "posttraumatic growth" refers to this quality more concretely: trauma and adversity can lead beyond recovery to greater psychological health.

Another kind of antifragility is called, usually pejoratively, the "Napoleon complex," based on the belief (for which, by the way, there is little evidence) that short people, like Napoleon, overcompensate for small stature by becoming overly aggressive or domineering, and it is just that compensatory trait that makes them successful. Napoleon himself wasn't especially short; at five feet, six inches tall (other sources say five-seven), he was in fact taller than the average adult Frenchman of his era.

The Antifragility of the "Contender"

Antifragility can be usefully applied to all kinds of competitions, including fighting. Professional boxers say that boxing is the sport in which even the winner gets badly beaten: if the winner doesn't take a few on the chin, it's not a fight; it's a massacre.

Some people seem to enjoy a good fight. A real contender doesn't mind taking some on the chin as long as he comes out on top; and the rougher the fight, the better he likes it and the harder he fights. A contender is not just impervious to being hit; it is taking blows as well as dishing them out that gives the fight its savor.

Contenders often have surprisingly mild personalities. They are not out looking for trouble. Their latent wildness and aggressiveness surface only once a fight has begun. I mentioned above how the crescendo of angry voices can be the harbinger of a barroom brawl. At that point, the mild, peacefully tippling world of Mildovia is about to give way to Wildovian chaos—unless, perhaps, the other party provides a mollifying counterargument. But such a peaceful resolution at this point in the proceedings is rare, even though a fight can almost always be avoided with a polite apology. It amazes me that people are willing to put their lives on the line to avoid losing face.

Once, as I was approaching a traffic light, I was daydreaming, and I braked too late and bumped into the car in front of me. It was a very gentle bump, and the only damage was a scuffed bumper. But the car was a shiny new BMW, and a muscular young man jumped out and repeatedly asked, "Why did you do it? *Why? Why? Why?*" He was obviously upset, and I could tell that the slightest provocation would lead to blows. I don't think he wanted to injure me seriously, although I would hardly have gotten away unscathed; rather, he just seemed to want to work off his anger. On a sudden inspiration, I replied, "Why? Because I fucked up. I'm truly sorry." The young man instantly relaxed. Fighting was now out of the question. In minutes we had exchanged information for the traffic report, all according to Mildovian rules of the road.

There are intellectual contenders as well. Recently, I gave a talk on the psychology of thought to a group of very good young chess players. I said that any of them who wanted to become great players had to love the game's combative aspects. You really have to be able to duke it out with your opponent across the chessboard. One of the team's coaches, a grandmaster—the highest ranking awarded by the World Chess Federation—though not a top-ranked international player, told me that what I said had given him an epiphany: "Now I understand why I never became a great player. I don't like to fight!" And sure enough, he was a chess player of an analytic, intellectual bent, always looking for the optimal move in any given situation, oblivious to the pugilistic aspects of competition. He nevertheless became a grandmaster because he usually succeeded in finding the best move.

But finding an objectively optimal move is not what fighting or competitive chess is about. A competition between two people is a struggle to find whatever works best at that moment against that opponent. Every great chess player experiences a chess match not as a struggle between two sets of chess pieces but as a battle between two human beings: two minds, two wills, two egos. A great chess player relishes trying to outwit the opponent. It is against the current opponent—his or her psychological makeup and playing style—that the player attacks and defends, while the objective "best move" does not have the last word. That is why in the great games, we might see a player trying a risky opening gambit or dangerous sacrifice, taking a few big blows that, if survived, might yield a stronger position.

I once asked some boxers, wrestlers, and those who don't engage in any form of physical combat who they thought would win if a boxer and a wrestler were to get into a fight. Let us suppose that both are first-class (but not world-class) athletes in their respective sports and of the same weight class. The overwhelming majority of the nonfighters said that the boxer would win, but both boxers and wrestlers voted unanimously for the wrestler. A wrestler might take a few punches as he moved in, but as soon as he got close enough that the boxer couldn't swing his arms freely, the boxer wouldn't stand a chance. The wrestler might, of course, get knocked out at the beginning, but that is likely only if the boxer is a world-class fighter. Most first-class wrestlers can take a few punches from a first-class boxer.

What about judo versus wrestling? When I asked judokas and Greco-Roman wrestlers who would win a fight between them, the answer was quite a surprise: I was told it would depend on what they were wearing. If they were clothed—for example in the traditional judogi, with its heavy jacket, lightweight trousers, and belt—then it is almost certain that the judoka would win, but a judoka is almost helpless against an almost naked body, since he will usually seize the clothes of his opponent, so if the opponents are scantily clad, the wrestler is highly favored.

Except for the occasional barroom brawl, boxers and wrestlers rarely fight each other in real life, but practitioners of chess come in a variety of combat types, and they clash across the chessboard every day. You might say that there are boxer types, wrestler types, and judoka types. And if you think the boxer types are all big, hulking fellows, think again. Judit Polgár, the best female chess player of all time, who used to be an international contender in the men's league, is an extremely aggressive player. She clearly enjoys intense competition, and she doesn't mind losing pieces. In fact, the more attacks she has to withstand, the stronger her game becomes. Like most big contenders, she is a gentle person away from the field of battle.

Many investors also fit the contender stereotype. In *What the Dog Saw,* Malcolm Gladwell compares two successful investors, the more traditional Victor Niederhoffer and our friend Taleb. Niederhoffer is a typical contender. He relishes competition and is willing to take enormous risks, but like a good poker player, he knows how to judge the competition and when to fold, so he typically makes large profits from day to day. Sometimes he fails to beat the odds and suffers a tremendous loss. On one disastrous day he lost his entire fortune, $130 million. He had to sell all his investments and his entire art collection, and when that wasn't enough to pay off his creditors, he had to mortgage his castle and even borrow from his children. He started again from scratch and soon regained his wealth, repurchased his art, paid off his mortgage, reimbursed his children, and became richer than ever. Niederhoffer hit rock bottom several times but always managed to recover. Taleb thinks Niederhoffer's investment strategy is a typically fragile one, but with Niederhoffer's contender attitude, it becomes antifragile, because it is precisely the huge losses that motivate him, and he is helped, of course, by the fact that the rich man's junk heap is a source of wealth. In this case, Niederhoffer's junk heap contained names of contacts in his Rolodex that he hadn't telephoned for years in favor of more important people, strategies that he knew but hadn't applied for decades, and knowledge of industries and individual companies that had been replaced by more efficient knowledge. After the crash, all this "junk heap" knowledge became useful once again.

Thus the rich man's junk heap is seen to be more than just a collection of objects that have been consigned to the dustbin or the attic. It also contains all the accumulated knowledge that has been put aside as obsolete or superseded. In a sense, mankind can be said to possess a junk heap containing a vast wealth of knowledge that has enabled us to change course after a catastrophe.

Taleb also has a typical contender attitude, but if Niederhoffer is a boxer, Taleb is more of a judoka, though from what Gladwell writes, he would likely have gotten into some brawls had his coworkers not held him back. As an investor, Taleb's fighting stance is based on the principles of antifragility: he never loses big, but he is willing to endure a large number of minor losses for a good chance at a big payday. Gladwell calls him "Wall Street's principal dissident," remarking that Taleb's early book *Fooled by Randomness* "is to conventional Wall Street wisdom approximately what Martin Luther's ninety-five theses were to the Catholic Church," or perhaps like someone suddenly using judo moves in the boxing ring.[2]

Investing as a Martial Art

Gödel's theorem has something to say about fighting. It tells us that there is no fighting style so powerful that no new fighting style can be invented to successfully oppose it. But then another style to counter that new one could be invented. This explains in part why so many forms of martial arts have evolved, with the result that everyone can find one that best suits his or her abilities and preferences. A world-class fighter in any martial art should be able to defeat a less accomplished practitioner of any other martial art in hand-to-hand combat, simply on the strength of being a better fighter. And a world-class fighter of any martial art will certainly defeat any garden-variety thug, even if attacked with a heavy bottle.

With all this talk about who can beat whom in a barroom brawl, it is worth mentioning that not every hostile encounter has to end in a knockout. Many years ago I answered a knock at the door of my university office and was greeted by a young woman who wanted to talk. She was the women's world champion in one of the lesser-known martial arts and had read in one of my books that a world-class grandmaster in any skill area has about ten thousand cognitive schemata in his or her head. She wanted to know whether the same was true of the martial arts. When I told her it assuredly was, she asked how she could know whether she possessed such knowledge. I replied that the fact that she was world champion proved that she did. She was very intelligent, and we had an interesting conversation, but I could not get her to understand my aversion to the martial arts and to fighting in general. "What would you do at night," she asked, "if you were attacked on the street?" I replied, "If someone came up from behind and knocked either you or me over the head with a baseball bat, none of our fighting skills—even yours—would be of any help. On the other hand, if I were face to face with my attacker and had a chance to speak with him, I would probably find a few words to dissuade him from hurting me, even if he took my wallet." The young woman thought about that and then said, "Well, I suppose that words are also a kind of weapon."

For most people, fighting belongs to the world of Wildovia and is far from their comfort zone. To win an intense competition, a fighter must draw strength from the blows received in combat. That is, antifragility is an essential feature of the successful combatant. Our friends the farmer and miller, whom we have encountered in this book from time to time, have no use for fighting. They are not running businesses that would benefit from the professional investor's combative attitude. They are happy when everything is running smoothly from day to day according to the rules of Mildovia. If they ever hit rock bottom, it would be very difficult for them to recover, and that is why they find it worthwhile to buy options against adverse markets even though they know that the investor selling the option is making a profit on their dime—though only while events are confined to their familiar and peaceful world of Mildovia.

Ever since options contracts came into existence, people have questioned whether options in general, or perhaps certain forms of them, should be banned. We saw above that laws restricting the options market date to the 1600s. A serious moral argument can be made in favor of restricting options trading, especially now that we know that it is theoretically impossible to put a proper price on some options contracts, namely, those that concern some of the thornier thickets of Wildovia. In Mildovia, the Black-Scholes formula makes it possible to set realistic prices for certain options, but no such formula exists—or can exist—in the wilder parts of Wildovia. Such options cannot be justified as "fair" in the usual sense of the word.

Yet whenever the regulations on options become too strict, there is always a huge outcry from every party involved. Farmers and millers complain because losing access to the options market would expose them to unnecessary risk. The options traders protest because they feel that the government should not be regulating their business. After all, they say, they won't come crying for help if things go wrong (though as we saw in 2008, that wasn't quite true). They will suffer the consequences and get back on their feet by themselves if they can.

It seems that investors believe in their own antifragility, even if Taleb thinks their modus operandi actually takes fragility to an extreme, since they risk sudden total bankruptcy. But investors, like others with a contender mentality, object to being forbidden to fight in the way they prefer. If dangerous sports like boxing are legal, why should the sport of investing not be legal as well? One answer to this argument might be that certain forms of investing pose a danger to a nation's economy. Such danger can be seen most clearly in investment strategies like Taleb's, because betting on the onset of a crisis can be a self-fulfilling prophecy, causing investor panic and market collapse. That is why, after the economic crisis of 2008, several countries passed legislation to regulate Taleb's style of financial martial art. We should note that all the martial arts have quite stringent regulations. Everyone knows that in boxing, for example, hitting below the belt is not allowed. Investors, like boxers, are generally willing to accept a certain degree of regulation, provided that they are not prevented from engaging in their chosen style of combat.

The ethical questions are more troubling. It has been said that one of the principal causes of the 2008 crisis was investor greed, and there is indeed some truth to that charge. Fighters very badly want to win, and in the heat of battle, they may not be particularly concerned about social niceties. They will use every legal means at their disposal, and perhaps even break the rules if they think they can get away with it. Greed—or if you prefer, the will to win—will never be extirpated. The question, then, is whether the investor greed that contributed to the 2008 crisis crossed into illegality—and if so, whether the laws should be more aggressively enforced or perhaps changed.

Our parents and teachers were no doubt correct in teaching us that fighting is not generally the most ethical way to settle disputes. But fighting, whether physical, intellectual, or financial, is not going to disappear. In any case, for those with a pugilistic bent, a fight is not primarily about settling a dispute. It is about the thrill of the fight itself. Banning fighting will not stop fighters from finding ways to fight, and there is the danger that such prohibitions, for all the ethical benefits they offer, might ultimately do more social harm than good.

The unadulterated antifragility represented by the boxer can benefit society—for example, when soldiers are needed. To take a less militaristic case, the pugnacious options trader who improves the lives of the farmer and the miller creates a social benefit for all of us. In a sense, investors are the white knights of Mildovia. We saw above that the spirit of Gödel's theorem can be applied to ethical and social issues, and we must therefore accept that the problem of balancing the ethical and practical implications of any particular legislation or regulation may not be solvable.

The White Knight of Mildovia

My childhood friend Alex, whom we've met in previous chapters, recently told me about a very interesting investment he was considering. He saw great promise in the company and thought it might even get listed on the Nasdaq, but he didn't quite trust the CEO. He felt somehow that the man did not have his heart set on making the company a global powerhouse. Alex worried that the CEO would be satisfied with making a sufficient profit for his own financial security and would then just sit back and coast. Even if Alex wound up with a decent profit, he would feel as if the CEO had simply stolen his money.

In the end, despite the company's promising prospectus, Alex decided not to jump in. He just didn't like the look of the CEO. Every venture capitalist worth his salt realizes that a startup's success depends as much on personalities as on the great new idea or business model. "A lot really depends on the CEO," Alex told me. "After all, there are no miracles." I had heard that line a hundred times, but this time it cut me to the quick. "So you think I'm writing a book about something that doesn't exist?" I asked. To my surprise, he answered, "Of course not. There's the miracle of LogMeIn, the first Hungarian company that made it to the Nasdaq." I pressed the point: "So you think that was a miracle?" He sighed and said, "Have you looked around here in Hungary? Can you imagine someone getting all the way to a major American stock exchange from this backwater? Only a miracle could have made it happen."

That wasn't just sour grapes. Alex was in no way trying to diminish the achievement of the founder of LogMeIn. It may have been a miracle, but without stubborn persistence and professional organization, no miracle would have taken the company all the way to the Nasdaq. One could say the same for the success of Rubik's Cube and the circumnavigation of the globe by the young English sailor Ellen MacArthur: if you haven't prepared meticulously, you won't be able to take advantage of the miracle when it appears. As Ernest Hemingway's old man put it in *The Old Man and the Sea*, "It is better to be lucky. But I would rather be exact. Then when luck comes you are ready."

The miraculous success of LogMeIn is what we have been calling a pseudomiracle. It follows from the nature of Wildovia that these sorts of things occur from time to time. The worldwide success of Rubik's Cube required both a pseudomiracle and a "true" miracle. First, that Ernő Rubik devised the cube's mechanism was a true miracle, since it was considered impossible according to the current state of science. And it was a pseudomiracle, as follows from the nature of Wildovia, that Tom Kremer was able to turn Rubik's toy into an international phenomenon.

My friend Alex also aims to create pseudomiracles. He looks for new enterprises that seem to have created a true miracle so that there is now the possibility of a positive black swan. But someone has to make it happen. That is the sort of battle Alex likes, and like every serious investor, he has a fighter's mentality. Although he used to provide venture capital to startups, that was not his principal contribution. It was working with management to turn a fledgling ugly duckling of a company into a first-class swan. Some people fight for a cause they believe in, even if they do not relish the fight, and they draw strength from their belief. That is not Alex. His antifragility is based on his fighter mentality. But he is not a fighter who will tilt at any windmill he happens to encounter. He needs a viable goal.

Alex continues not to believe in miracles, though he grudgingly admits that they exist, at least those we have called pseudomiracles and true miracles. But in talking about the CEO whose look he did not like, after he said, "There are no miracles," he added, "The leopard cannot change its spots." For the CEO to develop a new personality would, for Alex, constitute almost a transcendent miracle, and in those he absolutely does not believe. Whatever miracles he may have seen in the form of about-faces, conversions, and changes of heart have never been sufficient motivation for him to fight. He is an investor, not a psychologist or a pastor.

Alex is a very gentle person in private life, but in business, he can fight like a tiger. I have seen him do it often enough. But I am puzzled as to what type of fighter he is. He is certainly not a boxer, a wrestler, or a judoka—his style is quite different from all of them. If I were sufficiently versed in some of the other martial arts, perhaps I could determine whether Alex possesses the mentality of one of those or has invented one that is unique to himself. It is not out of the question that Alex has indeed invented a new martial art that he practices as an investor.

I would call Alex a first-class investor but not world-class. He doesn't pit himself against the likes of Niederhoffer, Buffett, Soros, or Taleb. His big dream, to grow a small Hungarian company to the point that it is listed on the Nasdaq, has never been realized, so he has never even made it into the major leagues. As soon as he realized that a startup for which he was working would not be a miraculous success, Alex would drop out. That is not to say that he made no contribution. Through his involvement, each company found itself on a stable footing with good prospects. All the stars seemed aligned for Alex to achieve a miracle through his stubborn persistence, just as they had been for Ernő Rubik and Ellen MacArthur. But something was missing. Perhaps it was just the luck of the draw or maybe just the fact that he is only a first-class investor, not world-class. Nevertheless, his efforts have saved many from the storms of Wildovia. I see Alex as a white knight of Mildovia.

The Advantages of Diversity

Above, I mentioned that a mathematical system cannot be just a little bit contradictory. It is either perfectly consistent or else inconsistent, and if it is inconsistent, you can deduce anything you like, including such nonsense as $1 = 0$. This suggests that mathematics is an extremely fragile structure, since a single obscure contradiction would cause the entire edifice to topple. So let us root for mathematical consistency. But even with consistency, a mathematical argument is a fragile thing, rendered useless by even a tiny logical error.

Every problem solved by mathematics—and even more so by science in general—and every new mathematical subject that is opened up raises new problems to be solved and new regions to explore. This mechanism makes science and mathematics robust enterprises, guaranteed not to come suddenly to a crashing end. Just as a fighter draws strength from the blows received, science draws new ideas and questions from those it has already addressed. It is in this sense antifragile.

The scientifically oriented chess grandmaster who doesn't like to fight, and so never became a truly great player, nonetheless managed over the years to knock out a few world-class grandmasters. How did he do it? Invariably it was by finding the objectively optimal move. In the games in which he got everything right, scientific analysis prevailed. But that kind of perfection is rare among human chess players, even grandmasters, because such scientific thinking is fragile. An antifragile fighter like Judit Polgár will deliberately not seek the objectively best move but try instead to create complexity, to strive for situations in which the best move is difficult to find, making it easier for her opponent to make a small mistake. Then she will pounce on that mistake, drawing strength from the fraught situation she has created.

The general antifragility of science is reflected in the fact that the best computer chess programs, which are far less subject to fragility than a single chess player, are able to defeat the best human players regardless of playing style. While our friend the scientifically minded grandmaster could defeat top players only occasionally, a chess program incorporating the thinking of hundreds of scientists can regularly defeat even the best warriors of the chessboard. The general antifragility of science can successfully counter the fragility of the individual mind. But in a tournament game with human players, the scientific player, with his fragile thinking, will usually be outmatched by the antifragility of the best fighters.

It may well be that the fighter mentality is a personal trait like extroversion in the sense that most of us possess it to a greater or lesser degree. But the nonfighters among us have other options for achieving antifragility, such as scientific thinking, which is fragile on the level of the individual but antifragile as an approach to problem solving.

Some scientists are fighters as well, but most are not. Science itself is antifragile not because its representatives are or because scientific thinking is, but because it realizes a different aspect of antifragility. Science works similarly to biological life, where antifragility comes from the fact that emerging species and constant competition for resources create an environment of robust diversity in which newer and ever more viable life forms can arise.

In chapter 4, "The Power of the Normal Distribution," we saw how biological life attempts to realize two seemingly contradictory principles: diversity and stability. It is the normal distribution that provides a good foundation for both of these at once, so it is not surprising that biological life is firmly established in Mildovia. But under Wildovian conditions, the diversity of biological organisms allows species not only to survive, but also to flourish. It is diversity that allows certain species to profit from environmental extremes. Among the diverse individuals of a population, there is a chance that some will find those extreme conditions to their liking. So diversity in itself can result in antifragility on the level of species. That antifragile structures can be constructed from fragile components is evolution's big idea. The same applies to science, where many scientists with fragile thinking are able to create distinctly antifragile scientific disciplines.

There may be other principles that can form the basis of antifragility, such as self-organization, but that discussion would take us too far afield, so I will just provide a few references in the notes for readers who wish to delve more deeply into the subject.[3]

Antifragility and Miracles

In Mildovia, we can more or less get a handle on adversity. Adverse effects have a mean and a standard deviation. They have a certain predictability and are not chaotic. We have an idea about how large an earthquake, how violent a storm, or how wide a lake we may need to confront, and it is sufficient that the systems we put in place be robust. They do not have to go beyond that to the point of being antifragile. If we can achieve such stability, we are prepared for almost every eventuality. In Mildovia, we can structure our world using the traditional tools of technology and economics.

In Wildovia, events are chaotic and unpredictable, and in its deeper thickets, phenomena don't even have a standard deviation. In such a world, planning, as it is generally understood, is impossible. Nobody can say what sort of preparations should be made for a catastrophe that is expected to occur perhaps once in a millennium. That requires fundamentally new planning principles that go beyond robustness and embrace antifragility.

Yet antifragility is not a panacea, even though it is an important concept for understanding certain principles, behaviors, and mentalities that are particularly useful in Wildovia. Before we can think clearly about how the concept of antifragility relates to Wildovian cataclysms, it would be useful to clarify what "antifragility" really means and perhaps try to find a more expressive name for it. If we are able to incorporate the concept of antifragility into our thinking, it should help us prepare for those black swans that inevitably appear in Wildovia from time to time. And if antifragility eventually becomes a well-founded scientific concept, we can expect to have better tools for handling those adverse pseudomiracles that result from the nature of Wildovia.

Adverse true miracles, meanwhile—those that cannot be explained by the current state of science—have a silver lining: it is through our response to them that science becomes antifragile, by drawing strength to evolve. For transcendent miracles, it is faith that provides an antifragile attitude, though the psychological mechanisms of that are currently obscure.

The Rich Junk Heap Yields Antifragility

So far, antifragility is the only general principle invented specifically for the conditions of Wildovia, although it can be useful in Mildovia as well. In the previous chapter, we considered four principles that do not meet the strict criteria of antifragility yet prove very useful under Wildovian circumstances while not contradicting the science of Mildovia (though they do not follow from it). Antifragility, though especially useful in Wildovia, also conforms well to a Mildovian mindset. Fighting, for example, exists in Mildovia as well as Wildovia, and it has been around since the dawn of time. The Law of Moses has something to say about it, in particular regarding proper retribution: "Breach for breach, eye for eye, tooth for tooth: as he hath caused a blemish in a man, so shall it be done to him *again*" (Leviticus 24:20). Combat can and should be regulated in both Mildovia and Wildovia, and antifragility is needed in both places. Gaining a deeper understanding of Wildovia with new tools may also help our everyday Mildovian thinking.

We have seen that both the miller and the farmer wisely let someone else take on the risks that they themselves are not cut out for. It is best to leave fighting to the fighters, just as we leave science to the scientists and ministering to the clergy. They are the white knights of Mildovia, and special rules have always applied to knights. Many envied the special status of the knights of old, yet few would have chosen to don their armor and engage in their style of combat. Extraordinarily large risks are best handled by specialists. Professional investors, due to their antifragility resulting from their fighter attitude, are better equipped than farmers or millers to take on large financial risks, so it is advisable to let them offer options contracts. They are more than willing do so, and under normal—Mildovian—circumstances, they usually become much richer than farmers or millers.

It is this wealth that allows investors to accumulate the junk heap that gives them the tools with which to recover after a major crash. If they cannot recover, it ought to be their own problem, but investors are only human, and they beg the government for outside help, allegedly on behalf of the farmer and the miller. This happened, for example, during the 2008 crisis. But Sweden, one of the few countries to ignore those cries, is also the country that best managed the crisis.[4] Experience since 2008 has shown that under such catastrophic circumstances, the public interest is best served by paying no heed to investors' personal interests. But this rule applies only to temporary measures; in the long run, if the country is hostile to investors, it is precisely the farmer and the miller who are ruined.

Occasional bankruptcy is simply a hazard of the investor's job (and the laws of Wildovia), and it is his or her sole responsibility. But when the laws of Mildovia are in effect, let us not envy investors' wealth or overregulate them to limit their capacity to accumulate it. If we did, there would be no one to assume the risks investors take, because the rest of us, lacking the antifragile investor mentality, could not stand the stress. The rich junk heap that the investors build in normal times contributes to the antifragility of the entire society.

13

Convertible Knowledge

Let us build a rich man's junk heap for our grandchildren rather than a poor man's house.

In 1757, Giacomo Girolamo Casanova, aged thirty-two, made a daring escape from the prison in the Palazzo Ducale in Venice, traveled to Paris, and set out to rebuild his life from scratch.[1] Luckily, an old friend of his was the foreign minister of France, and with his help, Casanova revived the French royal lottery. By turning his expertise in seduction to selling lottery tickets, he became a wealthy man.

But Casanova was not a successful businessman; he made and lost several fortunes, spending vast sums on his passion for seduction, which gained him more fame than had ever accrued to him from organizing a lottery. While trying to sell his lottery system to other countries, he managed to obtain audiences with the great rulers of the time, including George III of England, Frederick the Great of Prussia, and Catherine the Great of the Russian Empire, but they all rejected his proposals. It was not the age of great business thinkers.

Business Thinkers

The age of Casanova, and the following two or three centuries, were dominated by great innovators who were driven primarily by the spirit of innovation and only secondarily if at all by the business opportunities their innovations might open up. Most of today's thinkers about business, by contrast, are interested in solving scientific or technical problems only in order to create the Next Big Deal. An innovation is valuable only if it can be wrapped up in a narrative that diverts the flow of business in a new direction. Steve Jobs, for instance, started Apple Computer with the vision of creating a market for computers that people who were not technically inclined would buy and use. Over the years, he continued to look for ways to attract new users to his products. A more recent narrative was to make computing devices more portable: a computer in your knapsack, in your pocket, or on your wrist. For that, the bulky input device had to be eliminated, and for a long time, it seemed impossible that we would learn to control a small screen with our thick, clumsy fingers. But Jobs stubbornly stuck to his absurd idea, and thus were born the iPhone, the iPad, and now the Apple Watch.

Today almost every iPhone is made in China. But to what extent can the iPhone be considered a Chinese product if the manufacturer pays a 50 percent licensing fee to Apple, an American company, for every device it builds? Surely the business thinker here wasn't the person who decided that a Chinese company should manufacture iPhones, nor was it the person who organized production. It wasn't even the team that designed the iPhone itself. It was the person who had the intuition that the world would need a device like this and created the narrative of a universal information module in one's pocket. Steve Jobs was one of the world's greatest business thinkers.

A business thinker can't tell you today what he will be doing tomorrow, just as Casanova had no idea what he would do when he arrived in Paris. The only thing we can be sure of is that a business thinker will come up with a narrative that can divert the flow of business in a new direction. He or she may be a decision-making executive, an adviser to such executives, an entrepreneur, or a freelance intellectual.

It is the dreams of business thinkers and the narratives they create that form the new trends of production, and these trends represent a significant added value. Nowadays, it is fashionable to worry that manufacturing is being taken over by developing countries. What will happen to people in Europe and the United States without cutting-edge skills? The worst possible answer we can give is to train workers in those countries in specific skills, even if training includes a degree in teaching, engineering, or medicine. Obviously, teachers and doctors will always be needed, and (at least for now) needed locally. But in the developed world today, industries considered "intellectual-property intensive" account for about one-third of the gross domestic product, while construction and agriculture account for about 5 percent each. A larger and larger part of the gross global product—some estimate that it will reach 50 percent in a few decades—will be provided by business thinkers' visions of new gadgets, stuff, business strategies, and models—in short, narratives. That estimate may prove exaggerated, but the global race for economic leadership will be won by the companies, regions, and countries that have the best business thinkers.

Converting Knowledge

Perhaps unsurprisingly, business thinkers spend much of their time thinking, and much of that thinking is blue-sky dreaming. Which suggests that such thinkers might have backgrounds in philosophy, like George Soros; classical philology, like Charles Handy (a specialist in organizational behavior and management, widely considered one of the greatest business thinkers of our time); art history, like Esther Dyson (who became one of the first major Internet gurus); or other nonbusiness pursuits. It really doesn't matter what they studied; a conspicuously large number of the world's greatest business thinkers did not begin their intellectual lives thinking about business.[2] They trained in the humanities or in a theoretical science such as physics. The point is to make use of the broad knowledge and disciplined creative thinking acquired in a serious intellectual field, just as Casanova used his knowledge of human nature (which must have been considerable, given his record as a seducer) to make a success of the French lottery. Or just as Tom Kremer used his highly developed imagination to envision Rubik's Cube as a worldwide phenomenon.

You cannot learn business thinking from a book, just as you cannot become a successful sculptor or novelist from reading a manual. Whether you started out as a historian, philosopher, or astronomer matters little. If you hope to become a successful business thinker, you will have to convert your knowledge into something very different a number of times during your career, whenever the circumstances of Wildovia rise to the surface. Thus the most important goal of a twenty-first-century education is to gain a solid basis in some discipline and cultivate the ability to convert knowledge imaginatively as opportunities present themselves. A business thinker is a latter-day Casanova in the sense of being an intellectual adventurer, able to wield knowledge and imagination in a variety of situations, often in a field very different from that in which the knowledge was originally acquired.

Of course, business thinkers with far-roving minds cannot realize their ideas without a team of professionals to transmute grand visions into products, services, and business models. These pros also have to convert their knowledge, as the occasion requires, into something that can serve the business thinker's Wildovian dreams. The twenty-first century will be won by countries that not only provide the best business thinkers, but also line up behind them teams of trained professionals with profound and convertible expert knowledge. Their actual fields of knowledge do not matter nearly as much as the ability to convert it.

I would not advise any college student to choose a career path according to the demands of today's market. Who knows what the demand will be tomorrow? I would say instead, "It almost doesn't matter what you study. Just take it seriously, learn the interrelations of things and ideas; and however the world may turn, you will be in demand, because you will be able to apply your broad knowledge and skills to a variety of situations."

The Big Picture

We saw in chapter 9, "The Levels of Wildness," that innovations automatically amplify all four factors that lead to Wildovia. But we also saw that some people are robbed by innovation when the specialized knowledge that made them highly qualified and in demand loses its value. Their comparative advantage may disappear, and they may be forced to do something very different to earn a living. I usually tell both my psychology and engineering students that in twenty years, at least half, possibly three-fourths, of them will be doing something entirely different from what they are preparing for today.

What surprises me is that the twenty-year-olds of today are apparently not in the least surprised at this news. Even if they are unfamiliar with Wildovian science and the logic of miracles, they find it scarcely thinkable that they will be doing the same thing for forty or fifty years. This thinking applies even to those preparing to be teachers and doctors. Perhaps they will be teaching or healing all their lives, but with such different tools, under such different circumstances, and with such a different attitude that it can almost be considered a change of careers, and they will have to learn many things that they never contemplated while at university.

It follows from the nature of Wildovia that we often cannot predict what might compel us to change. Sometimes we cannot even imagine what surprises are in store for us. Nevertheless, we can prepare for the likelihood that we will have to change our professions radically and begin with something new, perhaps several times. That is why it is more important to possess a flexible worldview than concrete expertise.

Such was not always the case. Over the past few centuries, the motor of socioeconomic development, and the hallmark of individual success, has been specialized expert knowledge. In the twenty-first century, that role has increasingly been taken over by convertible knowledge. The ancients considered it a sign of unusual intelligence to be able to predict the future. In Steven Saylor's novel of ancient Rome, *Empire,* for instance, Claudius muses, "History, unlike divination, is an inexact science. This is because history deals with the past, which is gone forever and which neither gods nor men can alter or revisit. But divination deals with the present and the future, and the will of the gods, which has yet to be revealed. Divining *is* an exact science, provided the diviner has sufficient knowledge and skill."[3] In those days, teachers, who were considered less intelligent than soothsayers, were nonetheless thought to have a particular intelligence that allowed them to make the ideas of geniuses intelligible to a larger population. Today, a sign of intelligence is being able to prepare for a future that one knows is unpredictable.

Methods of teaching convertible knowledge are being formed today, but we are a long way from any concrete didactic principles. We can, however, already see that the teaching of particular trades is being taken over more and more by preparing students to understand things in a more general framework. There is a comprehensive viewpoint, a professional frame of mind, particular to every major field. Those who understand that viewpoint in depth have a good chance of internalizing the big picture, even during dramatic Wildovian situations, and that will help them get their bearings under a new and unexpected set of conditions.

Of course, we need to have something to convert. People must still thoroughly understand the science of Mildovia, but with less emphasis on the details and more on the big picture. Once, my father and uncle, both in their eighties, were brooding that they were losing their edge. At which point my uncle remarked, "We're lucky we had something to lose in the first place."

To create convertible knowledge, it is not the science and practice of Wildovia that we primarily have to teach. It suffices to understand a few general principles. There is no point in teaching about particular black swans if we are only explaining their occurrence after the fact. Knowledge of one black swan is no help in understanding the next and even less in predicting it. But it is precisely because of black swans that we need to teach the big picture of each profession. Only in this way can students be prepared, when a black swan appears, to oppose it with a consistent yet flexible professional mindset that can deal with radical changes in the world.

A famous professor of medicine used to hold a department meeting at the start of each semester to discuss the content of the coming term's lectures. A young assistant professor described how he had planned his lectures about renal function, leading to a fierce dispute among the kidney specialists in the meeting. After a while, the professor cut the dispute short by saying, "This is how kidneys will function this semester." He was not being cynical. What he meant was that although details of the lecturer's plan could be disputed, the course would give the students a good idea of how human organs work in general. Perhaps another plan could accomplish this as well. What was important was not which details of kidney function would be emphasized but that the students be exposed to a way of thinking that would help them grasp how medicine is practiced.

The idea that an understanding of the big picture can help in solving concrete problems is far from new. Ancient Egyptian priests used the mnemonic *method of loci* to memorize long lists, and since literacy was not widespread, memory played a much more important role in the preservation and transmission of knowledge than it does today. To use the method of loci, one imagines the items to be remembered next to various objects in a well-known place with a complex structure (such as a temple). The association of each object with its locus makes it possible to organize a large collection in a way that permits one to recall the items even after many years.[4]

The famous Russian psychologist Alexander Luria studied, over several decades, a man with an unusually strong memory. Once, Luria asked him to recall a long series of words that had been given to him fifteen years earlier. The man closed his eyes, paused, and said, "Yes, yes. . . . This was a series you gave me once when we were in your apartment. . . . You were sitting at the table and I in the rocking chair. . . . You were wearing a gray suit and looked at me like this. . . . Now, then, I can see you saying. . . ."[5] He then proceeded to cite flawlessly every word Luria had given him to memorize long ago. The man always associated the words to be memorized with the buildings and other details of Nevsky Prospect, Leningrad's (now St. Petersburg again) main thoroughfare. He was able to recall precisely how Nevsky Prospect looked at the time he had associated these particular words with its details. To accomplish his remarkable feat of memory, he, like the ancient Egyptian priests, needed a big picture that he knew exhaustively.

Teaching Convertible Knowledge

To teach convertible knowledge to students, we don't need to overturn the time-honored Mildovian methods of education. Students at the best schools have long received knowledge that is somewhat convertible, although the emphasis was elsewhere.

In an episode of the American television series *Northern Exposure*, Joel Fleischman, a New York City physician transplanted to a small town in Alaska, is flown by a local bush pilot to see a patient in a remote region. On the way back, their small plane malfunctions, and they crash-land in the vast northern forest. The pilot knows all about her airplane, but she is unable to locate the source of trouble. When she goes into the forest to gather food, the doctor takes a look at the engine. She returns to find Fleischman tampering with it, and she protests strenuously, to which the doctor replies, "It's only an engine, O'Connell. Actually, it's not unlike the human heart. . . . The thing is, the valve seemed stuck. It had this gunk on it like a pulmonary stenosis. . . . The heart has a valve. It gets stuck, and the blood doesn't move forward. It gets backed up, and then you have a major problem, hence, our situation here. . . . The thing was stuck. Now it's not stuck. . . . Come on. Give it a try." And to her astonishment, it works.[6]

The doctor has been forced by sudden Wildovian circumstances to apply his expertise in one area to something completely different. The big picture in his mind as to how certain things work proves to be convertible knowledge. Not because that was the intent of his medical school professors but because of the nature of high-quality knowledge.

While in medical school, the doctor learned the name of every groove, notch, and process of the temporal bone, of which there are more than two hundred. What was the point of that? There is no particular disease of the petrooccipital fissure or the sphenopalatine foramen, and even if there were, one could look up those names in an anatomy book or find them online in seconds. It is not, then, that each name is an important piece of medical knowledge. A medical student learns them because a physician needs to know every inch of the human anatomy. That is the basis of the doctor's expertise, his professional big picture. It is less important that a student acquire this or that particular big picture than that there be some such picture. So take your pick: delve deeply into Roman law, partial differential equations, or Indo-European linguistics.

The methods of professional training developed over the centuries have served well for the acquisition of usable knowledge, and a high level of knowledge in any field is always somewhat convertible. If you were to complain that making students memorize every detail of the temporal bone or the entire dictionary of Indo-European roots is only a sort of professional hazing, you would be half right. In a sense, such difficult, boring tasks are an initiation rite, but they are by no means pointless. They help a student form the general professional orientation that enables experts to comment intelligently on a wide variety of problems over a very wide domain, and they cannot do so without such an initiation. If we are constantly spending our time and energy looking up significant details, we won't have the mental space to invoke the big picture in our minds or even develop such a picture to begin with.

Nonetheless, the emphasis in professional training is about to change. There will be less focus on facts and details and more on understanding a multitude of expert models. Students will become acquainted with mutually contradictory models and so will come to understand that no model is universally applicable. One must always find the one that best suits the particular situation. Some of these models are Mildovian, while others are Wildovian. To be able to work with both kinds, students must have some familiarity with both Mildovian and Wildovian concepts. Above all, they have to learn to think in both ways.

Professional training in every discipline will continue to include what seem like unnecessary details, but those details will be more and more about gaining a deep understanding of a few fundamental models, some of which might never be used in practice. We will thus be able to teach convertible knowledge much more effectively than if we teach only the latest current methods, which tomorrow will be used only by the most backward practitioners.

The engineering students who attend my psychology courses find the following difficult to understand: If a man is walking down the street and a brick falls on his head, it might be partially his fault. He might have noticed and suppressed something suspicious about the building under which he was walking, or about the neighborhood, and so contributed to his bad luck. On the other hand, my psychology students need to be convinced that if a man walks down the street and a brick falls on his head, it might *not* be his fault. There is such a thing as plain bad luck. For psychology students, this randomness of the universe is hard to fit into their big picture.

Although we will continue to prepare students for particular professions, we must keep in mind that many students in both the sciences and humanities will find themselves working in such areas as marketing and finance, because that is where there are jobs for people able to comprehend and use complicated mathematical models and complex conceptual relationships. Scientific training focuses on the former, and training in humanities, on the latter. But a Wildovian singularity tomorrow may open up an entirely different line of work that needs those qualifications.

For most people, the professional "language" they acquire during their education remains with them throughout their lives, and they continue to view the world according to their chosen profession. Yet convertible knowledge that can be deployed under Wildovian circumstances is fundamentally different from traditional expert knowledge. Classical concepts such as optimization, minimization, and prediction are becoming less and less valuable, giving way to knowledge that can be applied to a variety of models, as we saw with the (fictional) doctor who fixed the airplane.

Models and Scale Models

When my daughter Vera worked as a fashion model, she used to joke that she was not actually a model but a scale model. A model, she argued, is something that works like the real thing but doesn't look like it, while a scale model is something that looks like the real thing but doesn't function like it, which is how she felt in the clothing she modeled—more like a dressmaker's mannequin than a real person.

There is knowledge in both models and scale models, but it is important to be able to determine which is needed in a particular situation. For applying convertible knowledge, both are necessary. A model is characterized by a domain of validity, a scale model by the big picture. The design of a business is usually a model, often called a business model. The impact assessment of a business idea is usually a scale model.

These two concepts are often blurred. When we construct a model of the DNA double helix, we are in fact building a scale model, because it does not work like real DNA; it just illustrates the big picture. Yet there is profound knowledge behind it, most obviously an understanding of the structure of the double helix itself—and that knowledge is the key to understanding how it works. Maxwell's equations, on the other hand, look nothing like the electromagnetic fields they describe, but both work the same way, though in different domains; thus the equations form a model of an electromagnetic field.

Whether you should employ a model or a scale model depends on how you want to use it. For example, I know Maxwell's equations quite well, and I know precisely that Maxwell's model says that electric current is present everywhere except in the wire. According to Maxwell, the wire's only role is to define the boundary of the field. Nevertheless, when my young children were about to touch an electrical outlet, I didn't cry out, "Don't touch it, because then you will become a part of the boundary of the electromagnetic field!" Instead, I used a scale model, and I cried, "Don't touch it, because the electric current will hit you!" This model seemed much more appropriate in that situation.

For today's decisions, especially for business decisions, gathering knowledge is no longer the issue it was; one can readily find both models and scale models online, though it is not always clear which is which. What's difficult is selecting the knowledge that is valid for the case at hand—in other words, validating knowledge.[7] We cannot know the general domain of validity for new pieces of knowledge, for new models and scale models. Moreover, their sources are seldom fully trustworthy. One of the biggest problems with the current state of the Worldwide Web is that the authentic, reliable sources of yesteryear, such as the *Encyclopaedia Britannica* and *Brockhaus Enzyklopädie,* have not adapted to the age of the Internet. Their online versions have not retained the authority of the printed versions. The Internet is too young for a trustworthy, generally accepted quality control system to have emerged, and trust obtained offline is not transferred automatically. The dominant philosophy of the Internet—total freedom of information as autonomous value—is inherently anti-authoritarian, even for websites with managing editors. And right now, the Internet is changing much too quickly for highly stable and trustworthy information sources to establish themselves.

For a long time, the Hungarian Wikipedia article on me contained the following "information": "His daughter, Réka Szabó, is a mathematician, an associate of Eötvös Loránd University (ELTE), leader of the independent dance ensemble Tünet Együttes."[8] But Réka Szabó works at the University of Technology and Economics, not ELTE, and she is also not my daughter. I did perform in her dance production *Chance* for ten years, though it was not my fancy footwork that landed me the role but because the production called for a mathematician to deliver a lecture on randomness.

Years ago at a party, I brought this up as an argument that Wikipedia cannot be trusted. The others at the party, all devout Wikipedians, took umbrage at my skepticism. But I noticed the next day that the article had been corrected. That is how Wikipedia works, so apparently randomly. We might say that it follows from Gödel's theorem that we cannot have trustworthy knowledge about the trustworthiness of the source of our knowledge.

I, too, use Wikipedia regularly, although I do not trust it fully, and I have also observed that the quality of its referenced sources varies. When a piece of information is important, I try to validate it by other sources. In chapter 7, "The Mathematics of the Unpredictable," for instance, I mentioned the story of Lavoisier's last experiment, which you can find on a variety of websites, though it is hardly more than an urban legend.

The Miracles of Our Grandchildren

Our grandchildren will continue to live their everyday lives in Mildovia, but the scale of occasional Wildovian pseudomiracles will become even broader. What we consider miracles now may become organic parts of their Mildovia, just as televisions, cell phones, heart catheters, cars, and air travel have become parts of ours, soon to be joined by genetic modification and renewable energy. As for the pseudomiracles that will be miracles for our grandchildren, not only can we not foresee them, but our grandchildren will not be able to predict them either. True miracles will also appear in our grandchildren's lives, but by then the boundaries of science will have shifted. Transcendent miracles, of course, will remain the stuff of myth and faith.

How can we help our grandchildren prepare themselves for the enormous negative black swans, the negative miracles of their era, that they are sure to face? One thing is certain: we cannot solve their problems in advance. To believe in such a possibility is simply delusional. The nature of Wildovia dictates that we can have no idea what burning problems will confront our grandchildren. What we now consider problems, such as global climate change, a giant asteroid impact, or the oil fields running dry, they may cite as proof of their grandparents' lack of imagination.

Since we cannot solve our grandchildren's Wildovian crises in advance, we can best help them by trying to make them as rich as possible. Let them have enough resources to train millions of experts with convertible knowledge in a wide variety of professions who will stand a good chance of solving whatever needs to be solved. Most important, let them have a junk heap rich enough to help them recover after a crash. We cannot prevent Wildovian crises from befalling them, but we can leave them models and scale models that have proved useful. Our current science is creating our grandchildren's junk heaps, both those of individuals and those of society as a whole.

We don't know what the future holds, but it is up to us to help determine what it can be. It is up to us to try to solve the problems of our own era: to master some of the deeper laws of Wildovia (such as determining the domain of validity of scale-invariance as a law of nature), to understand the chaotic nature of the subconscious mind, to develop the science of antifragility, to devise more effective methods of transmitting convertible knowledge—to mention just a few I have discussed in this book.

At the end of part I, I gave as an example of a true miracle the story of my cousin's sudden cure by means of a chamomile poultice. I did not tell this story to shame today's medical knowledge. Chamomile has been known since ancient times, yet even a few hundred years ago, almost everyone had a stubborn, irritating skin condition that was considered incurable. That this is no longer the case is due to medical progress, including both pharmaceuticals and (more significantly) the discovery of the importance of personal hygiene. My cousin's cure would not have been a miracle a hundred years ago, and it might not be one a hundred years from now, but according to the current state of science, it was a minor miracle. Our ancestors are to be thanked for that miracle, for it is they who made the junk heap of today's science rich enough to accomplish it. We, too, should prefer the rich man's junk heap to the poor man's house.

The miracles of the future cannot be predicted. There will be some that we find completely natural today and do not consider miracles at all. Others will be events we consider miracles now and will still consider miracles then, because although something may be unique and unrepeatable, it may recur, as unique and unrepeatable as it was the last time. Most important, there will be miracles that today we cannot even conceive. The logic of miracles is the logic of the unpredictable—a new kind of logic that we are just beginning to understand.

Epilogue

My personal miracle occurred in January 2006 in Berlin,
Germany.

I thought I knew Berlin, at least the eastern part. In my adoles-
cent years, I spent many summer vacations in the capital of what was
then the German Democratic Republic, and I loved the atmosphere
of the city, even as gray as it was in those days. Several years later, in
2006, I was walking the familiar streets of the reunited city when I
encountered a signpost: *To the Holocaust Memorial.* I was intrigued.
I hadn't heard of the controversies over its design and construction,
and I had no idea what it was.

I followed the signs until suddenly there were no more signs.
Where was the memorial? I went back to the last sign, followed the
arrow again. Still nothing. I was very surprised. That was not the
German exactitude I knew. Long ago I had stayed at a German sea-
side resort that had a sign on every street corner, with an arrow and
the words: *To the Sea.* I followed those arrows, and I was astounded
after the last turn when the unfathomable North Sea unfolded before me. There was one more signpost on the shore, which read:
The Sea.

But now in Berlin I saw nothing but a gray sea of concrete slabs.
I walked among them, and it took me some time before I realized
that I was in fact in the memorial: more than twenty-five hundred

pillars with identical rectangular bases and of various heights, less than eight inches high on the fringes, almost fifteen feet near the center (figure 23). It was like walking through one of those grim shoot-'em-up video games, where you fear to encounter something monstrous around any corner.

What impressed me most about the memorial was that it was right in the city center, occupying what was possibly Berlin's most valuable plot of land. The Brandenburg Gate was a few hundred yards away. An elegant luxury hotel was on the corner. Wim Wenders's angels soared above my head. The Germans had given this somber labyrinth at least four city blocks of their prime real estate.

Figure 23. The Holocaust Memorial in Berlin.

Then I gave in to other feelings. I was walking among this huge mass of stones as a descendant of Jewish ancestors, but that did not touch me. What overwhelmed me was the feeling that those who erected this monument were truly ashamed of what had occurred, of what they had done. I had to acknowledge the strength of mind and character it must have taken to erect this monument in the middle of their capital city sixty years after the events.

Then the miracle occurred. I began to feel love toward these people. I had always felt enormous respect toward those who gave us Bach, Goethe, and Gauss, but I had never thought I would someday come to love a nation that had exterminated the better part of my family. My father, who survived and lived to the ripe old age of eighty-seven, once said that our family didn't really know its propensity for longevity, since none of us had died of natural causes. That did not make me hate the Germans, but I thought there was no chance of my ever feeling love for them.

The fact that I did feel such love may be thought of as a transcendent miracle. It is at least a true miracle, since contemporary psychology has no explanation for the formation of such feelings. The phenomenon of romantic love can be understood psychologically to some extent, although I don't know of any mature scientific theories for it, so I am inclined to consider romantic love a true miracle as well. But if there is any scientific explanation for the total change of heart I experienced, I have yet to hear of it.

There is an underground museum in one corner of the memorial where all known exterminated Jews are listed, with German precision, one by one—over four million. That didn't move me either. It never even occurred to me to look for the names of my relatives. I returned to the concrete slabs and was taken once more by that strange feeling. This was not about my ancestors; the memorial was not about them, still less about their murderers. This memorial was about shame. It convinced me, not mentally but viscerally, that the Germans did not want to get beyond the horrors they had committed by pretending that time will heal all wounds or by denying or explaining the crimes away. This memorial exists to remind them, in the words that end Franz Kafka's *The Trial,* that "the shame should outlive them."

The memorial also represents a kind of antifragility, only the blow came not from an opponent but from the Germans themselves. And its form is another example of intelligent simplicity.

My experience offers yet another example—in addition to those of Ernő Rubik, Ellen MacArthur, and Harry Potter—that stubborn persistence and professional work and, we can now add, intelligent simplicity and antifragility can lead to a miracle. Thus it is no wonder that miracles have become an organic part of our world.

Notes

Chapter 1. On the Existence of Miracles

1. Quoted in Russo and Schoemaker (1990), p. 223. The authors note that while the story may be apocryphal, Watson is known to have said, "I do not believe in criticizing a man simply for making a mistake. If he shows that he has given the proper amount of thought to a matter, he shows that he has tried to do the right thing—and I am ready to forgive thoughtful mistakes" (ibid., p. 318).

2. See, for example, Gelderblom and Rouwenhorst (2005).

3. Gladwell (2009), p. 75n.

4. For more examples of Taleb's arrogance, see Taleb (2010), pp. xxxii, 37, 44, 274–276.

5. The poem quoted by Biberach and Luther's riposte: https://en.wikipedia.org/wiki/Martinus_von_Biberach.

6. Alfred, Lord Tennyson: "In Memoriam." https://rpo.library.utoronto.ca/poems/memoriam-h-h-obiit-mdcccxxxiii-all-133-poems.

7. Hofstadter (1979), p. xxi.

8. Ottlik (1966), p. 271.

9. Mullins and Kiley (2002).

10. Ottlik (1994), pp. 308–309, translation by Márton Moldován.

Chapter 2. The Mild World and the Wild World

1. See, for example, http://science.howstuffworks.com/life/inside-the-mind/human-brain/einsteins-brain.htm or http://en.wikipedia.org/wiki/Albert_Einstein%27s_brain.

2. For more on the precise formula for standard deviation and basic concepts of mathematical statistics, see, for example, Schervish (1998) and Shao (2008).

3. A distribution has other characteristics besides average and standard deviation; for instance, standard deviation will not tell you about the twin peaks of a bimodal distribution. But those are less frequently of real importance, whereas standard deviation is always fundamental.

4. After Mandelbrot and Hudson (2004), pp. 37–39.

5. On the Cauchy distribution, see Forbes et al. (2010) and Jondeau et al. (2007).

Chapter 3. The Source of Miracles: Gödel's Idea

1. Hofstadter (1979), p. 17, quotes the theorem in its original formulation.

2. On the philosophical background of Gödel's theorem, see Copi and Gould (1968). On the limits of mathematics, see Chaitin (2002).

3. Lem (1985).

4. Ibid., p. 194.

5. Borges (1988).

6. For a complete proof of Gödel's theorem, see, e.g., Hofstadter (1979), chapters 4–8, and Nagel and Newman (1983).

7. For example, the so-called continuum hypothesis proved to be Gödelian in the traditional axiomatic system of set theory (Cohen 1966).

8. Robinson (1996); Goldblatt (1998).

9. Robinson (1996).

10. See http://commenting-the-commentaries.blogspot.com/2007/05/johnny-von-neumann-jacob-bronowki.html.

11. Newton with autism: http://news.bbc.co.uk/2/hi/health/2988647.stm; with bipolar disorder: http://www.famousbipolarpeople.com/isaac-newton.html; with paranoia: http://www-history.mcs.st-and.ac.uk/Biographies/Newton.html.

Chapter 4. The Power of the Normal Distribution

1. A famous book of his in print today is Galton (1874/2008).

2. See https://en.wikipedia.org/wiki/Invictus.

3. See, for example, Mueller and Joshi (2000) and Doney and Mailer (2002).

4. See, for example, Adams (2009).

5. For information on biostatistics, see Lewis (1984).

Chapter 5. The Extremities of Mildovia

1. http://www.census.gov/hhes/www/cpstables/032011/hhinc/new06_000.htm.

2. On the central limit theorems, see Adams (2009).

3. Diamond and Saez (2011) and Simonovits (2015), for example, use a Pareto distribution of exponent 2 to model the distribution of incomes.

4. Koch (1999).

5. On the percentages of book sales, see Taleb (2010), pp. 235–236.

6. Forbes et al. (2010).

7. For Hilbert's list, see http://mathworld.wolfram.com/HilbertsProblems.html, http://en.wikipedia.org/wiki/Hilbert's_problems.

Chapter 6. The Sources of Equilibrium

1. The outlines of various proofs are presented at http://en.wikipedia.org/wiki/Brouwer_fixed-point_theorem.

2. Granas and Dugundji (2003); Border (1989).

3. Smith (1776/2014), book 1, chapter 2; book 4, chapter 2.

4. Arrow and Hahn (1971); Mandler (1999); Kornai (1971); Kornai constructs a theory taking the polar opposite position using the same mathematical apparatus.

5. Here are a few good references: Bodie et al. (2001); Kohn (2003); Mishkin (2001).

6. Dunbar (2000), pp. 100–108; Malkiel (2003); Bernstein (1998).

7. Black (1995).

8. See Mandelbrot and Hudson (2004), p. 64, and Lim et al. (2006), p. 67.

9. For a Black-Scholes calculator, see http://www.fintools.com/resources/online-calculators/options-calcs/options-calculator/.

10. Black (1989), p. 7.

11. Ibid.

12. Mandelbrot and Hudson (2004); Dunbar (2000), p. 93.

13. Mandelbrot (1999).

Chapter 7. The Mathematics of the Unpredictable

1. On hypnosis research, see Nash and Barnier (2012).

2. http://en.wikipedia.org/wiki/Butterfly_effect.

3. A popular science book on chaos theory is Gleick (2008); a more mathematically oriented treatment of the subject is given in Stewart (2002); on business applications, see Kotler and Caslione (2009).

4. Dietrich (2002); Zdenek (1993).

5. Molnár (2001); Damiani (2010).

6. Quoted in Dunbar (2000), p. 1.

Chapter 8. Scale-Invariance

1. Above are the one-week and one-hour graphs; below, the one-day and five-minute graphs. Graphs created from data supplied by Plus500.

2. A spectacular full-color animation used for figure 18 can be found at https://www.youtube.com/watch?v=zXTpASSd9xE#t=550.461.

3. The classic work on fractals is Mandelbrot (1983). On the history of fractals, see Mandelbrot (2002). Among the scholarly books are Falconer (2003), Schroeder (2009), and Sprott (1993). On pretty fractals, see Lesmoir-Gordon and Edney (2005), Peitgen and Richter (1984).

4. Pratt and Lambrou (2013); Freeman (1991); Hagerhall et al. (2004); Taylor et al. (2011).

5. See, e.g., Barabási (2002), Csermely (2009), and Palla et al. (2007).

6. On Lévy, see Mandelbrot and Hudson (2004), pp. 169–172.

7. Ibid., pp. 160, 161–162.

8. Barabási and Albert (1999).

9. The Matthew effect could also be called the Mark effect, because Mark 4:25 says the same thing: "For he that hath, to him shall be given: and he that hath not, from him shall be taken even that which he hath."

10. After Taleb (2010), chapters 14–17.

11. Granovetter (1973).

12. Csermely (2009), p. vii. Retranslated from the original Hungarian edition, p. 9. (This sentence was mistranslated in the published English translation.)

Chapter 9. The Levels of Wildness

1. After Mandelbrot and Hudson (2004), pp. 242–244.

2. Newman (2010).

3. The most frequent index is the Pareto exponent; see Adamic and Huberman (2002); Simonovits (2015); Newman (2005); Diamond and Saez (2011).

4. Lazear (1997).

5. On the Swiss referendum, see, e.g., http://www.wsj.com/articles/SB10001424052702304011304579217863967104606.

6. For those more sophisticated in mathematics, the Mandelbrot factor 1 corresponds to Pareto exponent 2.

7. Newman (2005).

8. The currently most commonly used model is VaR (value at risk); see, for example, Jorion (2006) and McNeil et al. (2005).

Chapter 10. Life in Wildovia

1. On the edge of chaos, see Cohn (2001) and Kauffman (1993).

2. Keynes (2000), p. 80.

3. Based on Taleb (2010).

4. Buckingham and Coffman (1999); Prahalad (2004).

5. For more on models using further normal distributions placed on the tail of normal distributions, see Novak (2011).

6. On the grand unified theory, see Hawking (1998) and Ellis (2002).

7. Orwell (2017), p. 34.

8. Ibid.

9. Ibid.

10. Keynes (2000), p. 80.

Chapter 11. Adapting to Wildovia

1. Taleb (2010), p. 203.

2. On the principle of comparative advantages, see Samuelson and Nordhaus (2009) and Heyne (2013).

3. Gladwell (2008), pp. 23–42.

4. Merton (1968).

5. On the Pygmalion effect, see Rosenthal and Jacobson (1968).

6. On whether such an exchange ever occurred, see http://en.wikiquote.org/wiki/Michael_Faraday#Disputed.

7. York (2009).

Chapter 12. Antifragility

1. For Taleb's hydra example, see Taleb (2012), p. 34.

2. Gladwell (2009), p. 56.

3. On self-organization, see Camazine (2003), Kauffman (1993), Bak (1996), and Haken (2010).

4. On an example in Sweden, see Soros (2008).

Chapter 13. Convertible Knowledge

1. Casanova (2001).

2. For the semiannual list of top fifty business thinkers, see http://www.thinkers50.com.

3. Saylor (2010), p. 48.

4. On the method of loci, see Mérő (1990), p. 107, or http://en.wikipedia.org/wiki/Mnemonic.

5. Luria (1987), p. 12.

6. *Northern Exposure,* season 3, episode 3 (s03 e03), "Oy, Wilderness."

7. Dörfler, Baracskai, and Velencei (2015); Iyengar (2011); Kotter (2012); Friedman (2005).

8. This can still be found in the history of the article.

Bibliography

Adamic, L. A., and B. A. Huberman. 2002. "Zipf's Law and the Internet." *Glottometrics* 3:143–150.

Adams, W. J. 2009. *The Life and Times of the Central Limit Theorem.* American Mathematical Society.

Anderson, C. 2008. *The Long Tail: Why the Future of Business Is Selling Less of More.* Hachette Books.

Arrow, K. J., and F. H. Hahn. 1971. *General Competitive Analysis.* Holden-Day.

Bak, P. 1996. *How Nature Works.* Copernicus.

Barabási A.-L. 2002. *Linked: How Everything Is Connected to Everything Else and What It Means.* Plume.

Barabási, A.-L., and R. Albert. 1999. "Emergence of Scaling in Random Networks." *Science* 286, no. 5439:509–512.

Bernstein, P. L. 1998. *Against the Gods: The Remarkable Story of Risk.* Wiley.

Black, F. 1989. "How We Came Up with the Option Formula." *Journal of Portfolio Management* (Winter).

———. 1995. *Exploring General Equilibrium.* MIT Press.

Bodie, Z., A. Kane, and A. J. Marcus. 2001. *Investments.* McGraw-Hill.

Border, K. C. 1989. *Fixed Point Theorems with Applications to Economics and Game Theory.* Cambridge University Press.

Borges, Jorge Luis. 1988. *Collected Fictions.* Tr. Andrew Hurley. Penguin.

Bronowski, J. 2011. *The Ascent of Man.* BBC Books.

Buchholz, T. G. 2008. *New Ideas from Dead Economists.* Plume.

Buckingham, M., and C. Coffman. 1999. *First, Break All the Rules.* Pocket Books.

Camazine, S. 2003. *Self-Organization in Biological Systems.* Princeton University Press.

Casanova, G. 2001. *The Story of My Life.* Penguin Classics.

Chaitin, G. 2002. *The Limits of Mathematics.* Springer.

Cohen, P. J. 1966. *Set Theory and the Continuum Hypothesis.* Addison-Wesley.

Cohn, N. 2001. *Cosmos, Chaos, and the World to Come.* Yale University Press.

Copi, I. M., and J. A. Gould. 1968. *Contemporary Readings in Logical Theory.* Macmillan.

Csermely, P. 2009. *Weak Links: The Universal Key to the Stability of Networks and Complex Systems.* Springer.

Damiani, G. 2010. "The Fractal Revolution." *Biology Forum* 103:151–190.

Diamond, P. A., and E. Saez. 2011. "The Case for a Progressive Tax: From Basic Research to Policy Prescriptions." *Journal of Economic Perspectives* 23, no. 4:165–190.

Dietrich, A. 2002: "Functional Neuroanatomy of Altered States of Consciousness." *Consciousness and Cognition* 12:231–256.

Doney, R. A., and R. A. Mailer. 2002. "Stability and Attraction to Normality for Lévy Processes at Zero and at Infinity." *Journal of Theoretical Probability* 15:751–792.

Dörfler, V., Z. Baracskai, and J. Velencei. 2015. "Mashup Content for Passionate Learners: Bridge between Formal and Informal Learning." In *Economics and Communication,* ed. M. Herzog, pp. 105–129. GITO. http://strathprints.strath.ac.uk/55423/.

Dunbar, N. 2000. *Inventing Money: The Story of Long-Term Capital Management and the Legends behind It.* Wiley.

Ellis, J. 2002. "Physics Gets Physical." *Nature* 415:957.

Falconer, K. 2003. *Fractal Geometry.* Wiley.

Ferguson, N. 2009. *The Ascent of Money: A Financial History of the World.* Penguin Books.

Forbes, C., M. Evans, N. Hastings, and B. Peacock. 2010. *Statistical Distributions*. Wiley.

Freeman, W. 1991. "The Physiology of Perception." *Scientific American* 272:78–85.

Friedman, T. 2005. *The World Is Flat: A Brief History of the Twenty-first Century*. Farrar, Straus and Giroux.

Galton, F. 1874/2008. *English Men of Science: Their Nature and Nurture*. Kessinger.

Gelderblom, O., and K. E. Rouwenhorst. 2005. "Amsterdam as the Cradle of Modern Futures Trading and Options Trading." In *The Origins of Value: The Financial Innovations That Created Modern Capital Markets*, ed. W. N. Goetzmann and K. E. Rouwenhorst. Oxford University Press.

Gladwell, M. 2008. *Outliers: The Story of Success*. Little, Brown.

———. 2009. *What the Dog Saw: And Other Adventures*. Little, Brown.

Gleick, J. 2008. *Chaos: Making a New Science*. Penguin Books.

Goldblatt, R. 1998. *Lectures on the Hyperreals*. Springer.

Granas, A., and J. Dugundji. 2003. *Fixed Point Theory*. Springer.

Granovetter, M. S. 1973. "The Strength of Weak Ties." *American Journal of Sociology* 78, no. 6:1360–1380.

Hagerhall, C. M., R. Purcell, and R. Taylor. 2004. "Fractal Dimension of Landscape Silhouette Outlines as a Predictor of Landscape Preference." *Journal of Environmental Psychology* 24:247–255.

Haken, H. 2010. *Information and Self-Organization*. Springer.

Hawking, S. 1998. *A Brief History of Time*. Bantam.

Heyne, P. 2013. *The Economic Way of Thinking*. Pearson.

Hofstadter, D. R. 1979. *Gödel, Escher, Bach: An Eternal Golden Braid*. Basic Books.

Iyengar, S. 2011. *The Art of Choosing*. Twelve.

Jondeau, E., S.-H. Poon, and M. Rockinger. 2007. *Financial Modeling under Non-Gaussian Distributions*. Springer.

Jorion, P. 2006. *Value at Risk: The New Benchmark for Managing Financial Risk*. McGraw-Hill.

Kauffman, S. A. 1993. *The Origins of Order.* Oxford University Press.

Keynes, J. M. 2000. *A Tract on Monetary Reform.* Prometheus. Originally published 1923.

Kindleberger, C. P. 2011. *Manias, Panics and Crashes.* Palgrave Macmillan.

Koch, R. 1999. *The 80/20 Principle.* Crown Business.

Kohn, M. 2003. *Financial Institutions and Markets.* Oxford University Press.

Kornai, J. 1971. *Anti-Equilibrium: On Economic Systems Theory and the Tasks of Research.* North-Holland.

Kotler, P., and J. A. Caslione. 2009. *Chaotics: The Business of Managing and Marketing in the Age of Turbulence.* AMACOM.

Kotter, J. 2012. *Leading Change.* Harvard Business Review Press.

Kun, M., and F. Szakács. 1997. *Az intelligencia mérése* (Measuring Intelligence). Akadémiai Kiadó.

Lazear, E. P. 1997. *Personnel Economics for Managers.* Wiley.

Lem, S. 1985. *The Cyberiad.* Tr. Michael Kandel. Harcourt, Brace.

Lesmoir-Gordon, L., and R. Edney. 2005. *Introducing Fractals: A Graphic Guide.* Icon Books.

Lewis, A. E. 1984. *Biostatistics.* Van Nostrand Reinhold.

Lim, T., A. Wen-Chuan Lo, R. C. Merton, and M. S. Scholes. 2006. *The Derivatives Sourcebook.* Now Publishers.

Luria, A. R. 1987. *The Mind of a Mnemonist.* Harvard.

MacRae, N. 1992. *John von Neumann.* Pantheon Books.

Malkiel, B. G. 2003. *A Random Walk down Wall Street.* W. W. Norton.

Mandelbrot, B. 1983. *The Fractal Geometry of Nature.* W. H. Freeman.

———. 1999. "A Multifractal Walk down Wall Street." *Scientific American* 280:70.

———. 2002. *A Maverick's Apprenticeship.* Imperial College Press.

Mandelbrot, B., and R. L. Hudson. 2004. *The (Mis)behavior of Markets.* Basic Books.

Mandler, M. 1999. *Dilemmas in Economic Theory.* Oxford University Press.

McNeil, A., R. Frey, and P. Embrechts. 2005. *Quantitative Risk Management.* Princeton University Press.

Mérő, L. 1990. *Ways of Thinking.* World Scientific.

———. 1998. *Moral Calculations.* Copernicus.

———. 2009. *Die Biologie des Geldes.* Rowohlt.

Merton, R. K. 1968. "The Matthew Effect in Science." *Science* 159:56–63.

Mishkin, F. S. 2001. *The Economics of Money, Banking, and Financial Markets.* Addison-Wesley.

Molnár M. 2001. "Low-Dimensional versus High-Dimensional Chaos in Brain Function—Is It an And/Or Issue?" *Behavioral and Brain Sciences* 24:823–824.

Mueller, L. D., and A. Joshi. 2000. *Stability in Model Populations.* Princeton University Press.

Mullins, G., and M. Kiley. 2002. "It's a PhD, Not a Nobel Prize: How Experienced Examiners Assess Research Theses." *Studies in Higher Education* 27, no. 4:369–386.

Nagel, E., and J. R. Newman. 1983. *Gödel's Proof.* New York University Press.

Nash, M., and A. Barnier. 2012. *The Oxford Handbook of Hypnosis: Theory, Research, and Practice.* Oxford University Press.

Newman, M. E. J. 2005. "Power Laws, Pareto Distributions and Zipf's Law." *Contemporary Physics* 46:323–351.

———. 2010. *Networks: An Introduction.* Oxford University Press.

Novak, S. Y. 2011. *Extreme Value Methods with Applications to Finance.* Chapman and Hall/CRC Press.

Orwell, George. 2017. *1984.* Houghton Mifflin Harcourt.

Ottlik, Géza. 1966. *School at the Frontier.* Harcourt, Brace and World.

———. 1994. "Két mese: Az utolsó mese" (Two Tales: The Last Tale), in *Hajnali háztetők, Minden megvan* (Rooftops at Dawn, Nothing Is Lost). Európa kiadó.

Palla G., A.-L. Barabási, and T. Vicsek. 2007. "Quantifying Social Group Evolution." *Nature* 446:664–667.

Peitgen, H.-O., and P. H. Richter. 1984. *The Beauty of Fractals.* Springer.

Prahalad, C. K. 2004. *The Fortune at the Bottom of the Pyramid: Eradicating Poverty through Profits.* Wharton School Publishing.

Pratt, G., and P. Lambrou. 2013. *Code to Joy.* HarperOne.

Robinson, A. 1996. *Non-Standard Analysis.* Princeton University Press.

Rosenthal, R., and L. Jacobson. 1968. *Pygmalion in the Classroom.* Holt, Rinehart and Winston.

Russo, J. E., and P. J. H. Schoemaker. 1990. *Decision Traps.* Simon and Schuster.

Samuelson, P. A., and W. D. Nordhaus. 2009. *Economics.* McGraw-Hill.

Saylor, S. 2010. *Empire: The Novel of Imperial Rome.* St. Martin's Press.

Schervish, M. J. 1998. *Theory of Statistics.* Springer.

Schroeder, M. 2009. *Fractals, Chaos, Power Laws.* Dover Publications.

Shao, J. 2008. *Mathematical Statistics.* Springer.

Simonovits, A. 2015. "Socially Optimal Contribution Rate and Cap in Proportional Pension Systems." *Portuguese Economic Journal* 14:45–63.

Smith, A. 1776/2014. *An Inquiry into the Nature and Causes of the Wealth of Nations.* Create Space.

Soros, G. 2008. *The Crash of 2008 and What It Means: The New Paradigm for Financial Markets.* Public Affairs.

Sprott, J. C. 1993. *Strange Attractors: Creating Patterns in Chaos.* M&T Books.

Stewart, I. 2002. *Does God Play Dice? The New Mathematics of Chaos.* Wiley-Blackwell.

Taleb, N. N. 2005. *Fooled by Randomness.* Penguin Books.

———. 2010. *The Black Swan: The Impact of the Highly Improbable.* Random House.

———. 2012. *Antifragile.* Random House.

Taylor, R., B. Spehar, P. Van Donkelaar, and C. M. Hagerhall. 2011. "Perceptual and Psychological Responses to Jackson Pollock's Fractals." *Frontiers in Human Neuroscience* 5:1–13.

Zdenek, C. C. 1993. *The Fractal Nature of Human Consciousness.* http://www.cejournal.org/GRD/zdenek.pdf

Index

Note: Figures and tables are indicated by "f" and "t," respectively, following page numbers.